Management for Professionals

More information about this series at http://www.springer.com/series/10101

Kenneth Moore

Measuring Productivity in Education and Not-for-Profits

With Tools and Examples in R

 Springer

Kenneth Moore
The Nous Group
Melbourne, VIC, Australia

ISSN 2192-8096   ISSN 2192-810X   (electronic)
Management for Professionals
ISBN 978-3-030-72967-7   ISBN 978-3-030-72965-3   (eBook)
https://doi.org/10.1007/978-3-030-72965-3

© The Editor(s) (if applicable) and The Author(s), under exclusive license to Springer Nature Switzerland AG 2021
This work is subject to copyright. All rights are solely and exclusively licensed by the Publisher, whether the whole or part of the material is concerned, specifically the rights of translation, reprinting, reuse of illustrations, recitation, broadcasting, reproduction on microfilms or in any other physical way, and transmission or information storage and retrieval, electronic adaptation, computer software, or by similar or dissimilar methodology now known or hereafter developed.
The use of general descriptive names, registered names, trademarks, service marks, etc. in this publication does not imply, even in the absence of a specific statement, that such names are exempt from the relevant protective laws and regulations and therefore free for general use.
The publisher, the authors, and the editors are safe to assume that the advice and information in this book are believed to be true and accurate at the date of publication. Neither the publisher nor the authors or the editors give a warranty, expressed or implied, with respect to the material contained herein or for any errors or omissions that may have been made. The publisher remains neutral with regard to jurisdictional claims in published maps and institutional affiliations.

This Springer imprint is published by the registered company Springer Nature Switzerland AG
The registered company address is: Gewerbestrasse 11, 6330 Cham, Switzerland

Preface

This book exists because of the challenges I faced during my transition from PhD student to consultant. Those challenges represented a personal test of shifting focus from knowledge creation to translating knowledge into practice. I quickly had to become acquainted with, not only a new paradigm for guiding my work, but also a new set of technical skills to facilitate that work. I quickly realised a gap between research in my field and how it is used in practice to help institutional leaders make decisions on the ground.

This book attempts to build a bridge between advanced analytical techniques being developed in the research space and their applications for helping leaders and managers better run their organisations.

My PhD research fell within the broader field of education studies. Education research is often geared toward producing findings that are useful for education practitioners and teachers. Findings are generally positioned either directly or indirectly to facilitate the delivery of higher quality education. The education research space, however, is dominated by qualitative approaches. In general, this is no mistake. Driving education policy reforms with nothing but statistics or performance metrics may not be advisable if you are after an equitable education system.

My dissertation explored appropriate uses of performance measurement in context, the creation of new metrics, and the policy and management implications of putting them to use. However, I am not here to tell you that education practitioners should blindly follow mathematical models of education institutions. What I am arguing, though, is that in this era of abundant data and skyrocketing computational capacity, we cannot afford to overlook opportunities to learn from the data we store and the evidence we collect.

Prior to my PhD research, I worked as a program evaluator for non-for-profit organisations (NFPs). And during my PhD candidacy, I was constantly making connections between the the techniques I was developing for education and their applications for NFPs. My prime focus was to develop appropriate productivity metrics for organisations whose *raison d'être* is not to turn a profit. But PhD dissertations must remain narrow in focus in order to ever complete them! This book now presents examples for how the techniques I was developing for higher education institutions can be applied to NFPs as well.

This book was also written during the 2020 COVID-19 pandemic. Our world is facing rapid and accelerating change, and it is going to be difficult to judge for quite some time how the decisions we are making now are going to affect us years down the road. Some industries have exploded with success during the pandemic, but others have suffered. For education and not-for-profits it will be increasingly important to keep a finger on the pulse of our organisations. The techniques in the book are not intended to serve as levers for performance management, but rather as new tools for performance improvement. I intend to make these techniques more accessible for both junior and experienced analysts to inform their leaders of emerging opportunities and risks for their organisations.

At the end of the day, I hope that I can shed some clearer light on one part of the academic literature that has been a large part of my life for several years.

Thank you!

Melbourne, VIC, Australia   Kenneth Moore

Contents

Introduction

1.1 The Value of This Book

This book demonstrates how to measure productivity for organizations whose reason for being transcends the objective of earning profits or maximizing margin. By the end of the book, you will have a grasp on several common techniques for measuring productivity. You will learn how to leverage these techniques' strengths for the education and not-for-profit (NFP) contexts, and you will learn where to avoid the pitfalls of conducting a quantitative analysis for organizations that are often driven more by values than by numbers.

You will build a hands-on understanding of sophisticated measurement techniques commonly used by professional economists and researchers. And you will emerge from the text with a set of replicable and adaptable tools that can be applied to your own data and contexts.

The book is designed to make the impenetrable accessible. We avoid mathematical notation and jargon where at all possible. Instead, we divert directly to usable scripts that can be input directly into your console, so you can see exactly what's happening without having to wade through technical language and descriptions.

This book gives you the ammunition to conduct powerful analyses and trains your restraint in taking them too far. While this book minimizes the time it spends on argumentation, it does walk you through essential contextual considerations—learned from experience. Those who hold the greatest stake in the output of many of the analyses showcased in this book are often not as moved by performance metrics as they are by principle. More than anything, this book's examples strive to avoid the pitfall of 'lies, damn lies, and statistics'.

Together, we'll work through examples with the dual purpose of arriving at results that are both robust and persuasive. Anyone who has tried to put a few extra decimal places on a number to try and influence a decision regarding a low-performing school district likely knows what it feels like to have eyes roll at their

© The Author(s), under exclusive license to Springer Nature
Switzerland AG 2021
K. Moore, *Measuring Productivity in Education and Not-for-Profits*,
Management for Professionals, https://doi.org/10.1007/978-3-030-72965-3_1

presentation. In this book, while we do aim for precise and reliable findings, we also know that the true value of conducting a productivity analysis for an education institution or an NFP is not the number you arrive at—rather, it's what you learn from the journey of conducting a rigorous measurement exercise.

1.2   Who Is This Book For?

This book targets those who maintain a strict commitment to NFPs staying true to their missions, but who acknowledge that all budgets and strategic plans are drafted under the constraint of scarce resources. The book is not intended to provide tools for performance management, but rather position an organization for performance improvement. Those who might be interested in the book are burgeoning data scientists, consultants, BI analysts, program evaluators, and even the technically inclined organizational leader or executive.

1.3   Background and Motivation

This book exists because of a commitment to reproducible research. The techniques shared here were initially developed for a PhD dissertation.[1] This book takes the content out of the academic literature and refines it for a professional audience. The methods we cover here are no longer meant for contemplation but for implementation.

The world of productivity measurement is dominated by economists, engineers, and operations researchers. Most of their techniques have been developed for circumstances where the precision of results is as important as their accuracy. The list might include micro and macro economies, manufacturing plants, hospitals, or other technological industries. As researchers and education economists have employed such sophisticated methods for NFPs, schools, and universities, taking a step back to first principles, reengineering niche methods, and ensuring that adaptations appropriately fit new contexts has been a surprisingly uncommon occurrence.

This book intends to unpack a great deal of complexity around the practice of productivity measurement for education institutions and NFPs, demystify some common methods, and repurpose their use for new circumstances and new audiences.

1.4   Assumptions

This book was pieced together under the assumption that you, the reader, want to learn how to measure productivity for education institutions and NFPs. There exist many ongoing arguments about whether or not productivity measurement in

---

[1] Moore. K. (2019). Investigating higher education productivity and its measurement in Australia. Melbourne Graduate School of Education. Available at https://minerva-access.unimelb.edu.au/handle/11343/238674. Accessed 30 Nov. 2020

education is a good thing or a bad thing, what techniques might be better suited than others, or which indicators should or shouldn't be considered. Many sides of these arguments are entertained in the dissertation mentioned above, but they will not be rehashed here. We'll get right into the meat and discuss strengths and limitations of different measures.

A grander assumption at play here is that we move forward with a shared understanding that "productivity" is one aspect of "performance." An organization's performance may hinge on culture, politics, relationships, external regulations, and policy. All of which are important and must be taken into consideration for a full performance assessment of any organization or institution. Productivity is an objective measurement of where the rubber hits the road with respect to internal work processes and procedures. Productivity is concerned with measurable inputs and demonstrable achievements. The result of a productivity measurement exercise is an objective benchmark against which gaps can be quantified. Myriad factors may influence the size of a measured gap, or even whether a measured gap should be considered something to overcome. Contextual factors can and should color the analysis. By itself, the productivity indicator serves as an objective anchor point for conversation, debate, and further investigation.

A final assumption is that this book presumes the reader has some experience with the R programming language and using RStudio. Examples in this book do start simple, and the second half of this chapter serves as an introduction to some common functions and packages we'll be using throughout the book. If you are an absolute beginner to R, however, it's advisable that you take at least a little time to familiarize yourself with basic R commands and functionality, RStudio and its interface, and how to set up a project and/or working directory. Myriad resources for this exist online for free and at modest prices.[23] Two recommended resources include Hands-On Programming with R and R for Data Science.

1.5   Why R?

There comes a time in most analysts' careers where spreadsheets (or really any other point-and-click interface) become too cumbersome to perform increasingly complex operations on increasingly large sets of data. The author reached that point 6 months into his PhD program, and if you're reading this book, you likely don't need any further convincing of the benefits of transitioning to one of the core vehicles for data science, such as R or Python.

There is no fundamental reason why we choose R over Python. Both languages are excellent, and neither will be going away any time soon. A few quick Google searches will bring you to endless debates over which is better for what purposes. But the more basic fact is that advances in one language often propel advances in

---

[2] Grolemund, G. (2014). Hands-on Programming with R. O'Reilly Media, Sebastopol

[3] Wickham, H. & Grolemund, G. (2017). R for Data Science. O'Reilly Media, Sebastopol

the other, and both have large communities of developers and data scientists constantly updating and maintaining their functionality.

One compelling reason why we choose R in this text is the Tidyverse,[4] a collection of packages designed for data science, which share a common design philosophy, grammar, and data structures. Some R-purists will raise concerns with the Tidyverse-style of programming. No aspect of data science is immune to debate! However, the sentiment of the author and his closest colleagues—and one you can also find repeatedly online—is how effective the Tidyverse is at opening up data science and computer programming to relative beginners. And there is almost no end to its utility no matter how advanced you get.

R's Tidyverse is the prime reason why this book can exist in its current form. After applying the philosophies and practices of "tidy data" to thousands of lines of code that formed the basis for an entire PhD analysis, I have been able to reduce much of it to a few concise chapters with commentary. And the content of which is ready to be implemented and replicated by anyone who is technically inclined and with no more than a bachelor's degree.

1.6   Terminology and Contextual Considerations

In the case of education and NFPs, the term "productivity" often does not fall well on the ears of the people who are doing the work. Referring to resources and accomplishments as inputs and outputs may serve to objectify key human elements of the services being delivered. Any analyst must be fully aware of this reality because colloquialisms have the potential to destroy the credibility of an otherwise good analysis.

For example, in a school to refer to all students as "inputs" grossly reduces the nuance and complexity that a teacher must consider when delivering a lesson to a full classroom. At the same time, if we compare two similarly sized schools thought to be facing similar environmental and pedagogical challenges, and we find that the have vastly different student progression rates, then the cold hard numbers can serve as the impetus for challenging assumptions and directing further investigation.

An important context to understand is that performance management and performance measurement regimes in education and NFPs have given rise to legitimate concerns about measurement practices producing more deleterious effects than benefits. An adage to keep in mind is "Goodhart's Law": once a measure becomes a target, it ceases to be a good measure.

The pendulum, however, can swing too far in either direction, and you may encounter colleagues or practitioners in education or NFPs who will dismiss a piece of evidence largely based on its being presented in numerical form. Such a standpoint represents only opportunities lost. The fact is that operational and management decisions will be made with or without good information and data. And if we

---

[4] Tidyverse. tidyverse.org

train our leaders and analysts to be mindful of context with quantitative results, to not compromise on rigor, and to proceed with caution, then more opportunities will arise to use abundant data already available and powerful analytical tools to much greater effect.

1.7 Structure of the Book

1.7.1 Chapter Structure

Each chapter of this book is structured with a conceptual and contextual introduction. The chapters then proceed to a tutorial for applying the concepts and techniques introduced. Each chapter then concludes with a reflection section that raises considerations and caveats for implementing the demonstrated techniques in a real-world context.

1.7.2 Chapter Progression

Chapters 2 and 3 serve as an introduction to the core concepts, philosophies, and methods of productivity measurement. The tutorials focus on how a productivity measurement exercise could be carried out in a secondary education setting with two different school districts. The objective of these chapters is to measure productivity from multiple angles to represent different interests in performance and capture a range of strengths and weaknesses of the two school districts under investigation.

Chapters 4 and 5 shift focus to the setting of higher education and consider data from four different universities. The purpose of these chapters is to demonstrate how productivity can be measured when consistent standards or methods simply do not exist to quantify the value of outputs. That is, when you can't place a precise dollar figure on the value of university degree or a research publication, how can you generate a meaningful and helpful productivity indicator, which can serve as an agreeable benchmark across a field of institutions? We address that question with a novel application of a technique, called the Törnqvist Index (TI), which estimates productivity change over time.

Chapters 6 and 7 shift to a not-for-profit setting. The tutorial presents a government agency that has contracted six service providers to deliver a new program across six different locations within a state. Once again, on a fixed budget when margin does not indicate how well these service providers have performed, what data should be considered, and how should it be treated, to increase understanding about which of these organizations' strategies have been most effective? We unpack a technique called data envelopment analysis (DEA), build our analysis from the ground up, and offer a unique application of this method to uncover some deeply hidden but consequential trends in the data.

Chapters 8 and 9 can be thought of as bonus chapters, where we go back to our higher education data and further adapt our TI methods, based on some learners from the DEA analysis with the NFPs. We also learn how to bolster our findings by testing some hypotheses with a time series analysis. Stakeholders will often want to know the extent to which they can trust results, and a well-structured test for statistical significance can add credibility to results.

1.8   Getting the Most Out of this Book

1. *Play along.* This book is designed to be used. You will get far more out of this book if you have an R session open on your computer as you work through the tutorials. Each script was designed to be fully reproducible. None of the tutorials ask you to download data from a website that might crash or change its URL in the next 6 months. All mock data are created in line with the script.

2. *Experiment.* As the chapters progress, some of the scripts become more complex. Break them up if necessary. If you run the scripts from the top, verbatim, they will work, but results may still sometimes surprise you. Even though the code has been commented out in chunks—in addition to the expository notes leading you through the tutorial—some operations are inevitably complicated. In these cases, once you have pasted the code into your R session, you can run smaller chunks at a time, so that intermediate steps become more explicit. You can further interact with several examples in this book at https://doi.org/ikenresearch.com/interact.

3. *Note the dependencies.* If you are new to productivity measurement, it's best to work through this book from beginning to end. If on the other hand, you have heard of TI or DEA before, and you just want to see how they work and how to implement them, you can skip to those chapters. Just note that some of the tutorial scripts depend upon data or functions created from previous tutorials. Each dependency is as follows:
   - Chapter 3 requires Chapter 2
   - Chapter 5 requires Chapter 4
   - Chapter 6 requires Chapter 7
   - Chapter 8 requires Chapter 5
   - Chapter 9 requires Chapter 4

4. *Use all resources at hand.* When in doubt about how an R function works, you can always use R's built-in '?' function to display information about the function and to get examples of how to use it. R also includes several vignettes that provide additional walk-throughs, so users have many options for testing packages and functions. To see a list of all the vignettes you have available, just enter *vignette()* into your console. Entering *vignette('dplyr')* into your console, for example, will bring up a full introduction to that package with workable examples in your RStudio's "Help" window. The accompanying website for this book further allows you to play with examples and dynamically manipulate data.

1.9   Tutorial: A Brief Intro to Some Tidyverse Functions

1.9.1   Setup

Each tutorial in this book begins with a "Setup" section that tells you what packages you need to activate (or install if you haven't already) in order to run the code in the script. If you don't have the Tidyverse yet, for example, you need to install it with the *install.packages()* function first. Sometimes the setup section will ask you to load data created in previous chapter as well. Our current tutorial requires that you activate only the four packages below.

```
# functions for tidy data management
library(tidyverse)
# create simple formatted tables
library(knitr)
# intuitive for working with text
library(glue)
# formatting scales and data labels in ggplot charts
library(scales)
```

A final note on this tutorial is that you will need to paste any code shown here into your console or scripted editor to see the outputs. In the chapters that follow, we will examine intermediate outputs of our analysis more closely and view them in tables, so we can appropriately interpret results as they come in. For this first brief introductory tutorial, however, we just demo the use of some functions that we'll be using frequently throughout the book, and as the reader, you may engage with it to the extent that you feel necessary.

1.9.2   Intro to the "Pipe"

The pipe function helps you compose more intuitive code that's easier to create and easier for others to follow. When you know you need to perform multiple operations on a single variable or object in sequence, the pipe lets you write the code exactly in that sequence, rather than as a complex statement of nested brackets. For example, consider three equivalent chunks of code below, starting with the base R syntax and ending with tidy piped syntax.

```
# base R
x <- seq(sum(rep(abs(-1),10)), to = 20, by = 2)
# piped
x <- -1 %>% abs() %>% rep(10) %>% sum() %>% seq(to = 20, by = 2)

# tidy piped
x <- -1 %>%
```

```
abs() %>%
rep(10) %>%
sum() %>%
seq(to = 20, by = 2)
```

1.9.3   Using Filter, Select, Mutate, Group, and Summarise

The *filter()*, *select()*, *mutate()*, *group()*, and *summarise()* functions are all included in the *dplyr* package. As stated above, a full *dplyr* tutorial is available with *vignette('dplyr')* once you have installed the Tidyverse.

To make use of these functions, we need a data frame to manipulate. Use *tibble()* to create a new data frame. The *tibble()* function is also included in the Tidyverse and has several advantages over using base R's *data.frame()* function.

```
# create data frame
sample_data <- tibble(
  type = rep(c("A", "B", "C"), 4),
  category = c(rep("X", 6), rep("Y", 6)),
  value = seq(1, 12, 1)
)
```

Select certain columns from the data frame and filter for specified rows.

```
# select only the the type column
sample_data %>%
  select(type)
# exclude the category column
sample_data %>%
  select(-category)
# include only category X observations
sample_data %>%
  filter(category == "X")
# include type A and B observations
sample_data %>%
  filter(type %in% c("A", "B"))
# exclude category X observations
sample_data %>%
  filter(category != "X")
# exclude type A and B observations
data %>%
  filter(!type %in% c("A", "B"))
```

Now use *mutate()* to create some new rows.

```
sample_data %>%
  mutate(fraction = value/2,                    # value column
divided by 2
          prop = value/sum(value),              # proportions of total
          scaled = value/max(value),            # normalise values with
respect to max
          cat_2 = rep("1.a", n()),              # new column with all
"1.a" values
          cat_3 = glue("{type}|{category}"),    # new column with type
and category
          value = as.character(value))          # transform into a
character column
```

Separate the data into groups, and perform operations individually on each group.

```
sample_data %>%
  group_by(category) %>%
          # find proportionate values for cats X and Y separately
  mutate(cat_prop = value/sum(value),
          # normalise values with respect to the max value
          # for cats X and Y separately
          cat_scaled = value/max(value)) %>%
  ungroup()
```

Group data and summarize.

```
# find summary figures for categories types A, B and C separately
sample_data %>%
  group_by(type) %>%
  summarise(total = sum(value),
          max = max(value),
          mean = mean(value),
          sd = sd(value)) %>%
  ungroup()
```

1.9.4   Long and Wide Format

We'll frequently use the *tidyr* package to reshape data from long format to wide format and vice versa depending on what types of calculations we need to make. The prime functions we'll use to accomplish this are *pivot_longer()* and *pivot_wider()*. For a more complete demonstration of these functions, you may refer to *vignette('pivot')*.

```
# summarise for category and type to get
# unique observations for each combination
data_unique <- sample_data %>%
  group_by(type, category) %>%
  summarise(value = sum(value)) %>%
  ungroup()
# pivot to wide format, where all types
# appear as their own columns
data_wide <- data_unique %>%
  pivot_wider(names_from = "type", values_from = "value")
# pivot back to long format
sample_data <- data_wide %>%
  pivot_longer(A:C, names_to = "type", values_to = "value")
```

1.9.5   Using ggplot2 for Visualization

The *ggplot2* package is the gold standard for creating visualizations in R. There is a learning curve associated with its use, but it becomes intuitive once you get the hang of it. There are many tips and tricks and myriad resources available online for this package. The key advice we'll highlight here is that your data should be in long format before attempting to plot with ggplot. Long format is not a hard requirement, but several common applications of *ggplot2* depend upon data being in long format. Below we'll demo some standard good practice for using ggplot. Adding new layers to a plot or tweaking elements of the plot must be specified with '+' once you've called the *ggplot()* function. If you go to your "Help" menu and hover over "Cheatsheets," you will find a fantastic guide for all sorts of ggplot charts and functions.

Throughout the book, you'll see increasingly advanced applications of *ggplot2*, where we manipulate the axes, edit data labels, and explore different plotting options with the different geometries available. For this intro, however, we'll keep things very simple.

Plot a column chart (Fig. 1.1).

```
# prep data for ggplot
d <- data_unique %>%
  group_by(type) %>%
  summarise(value = mean(value)) %>%
  ungroup()
#plot
d %>%
  # set key aesthetics (axes)
  ggplot(aes(x = type, y = value)) +
  # chose chart type
  geom_col()
```

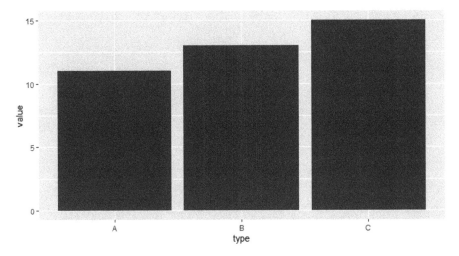

Fig. 1.1   Column chart example with ggplot2

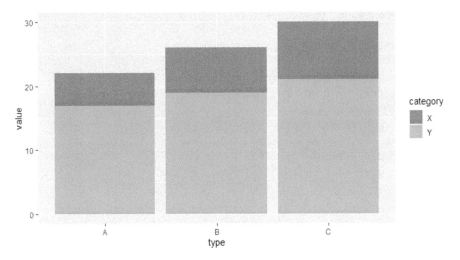

Fig. 1.2   Stacked column chart example

Plot a stacked column chart (Fig. 1.2).

```
# plot straight from data_unique
data_unique %>%
  # set key aesthetics (axes and stacking groups)
  ggplot(aes(x = type, y = value, fill = category)) +
  # chose chart type, and specify stacked
  geom_col(position = "stack")
```

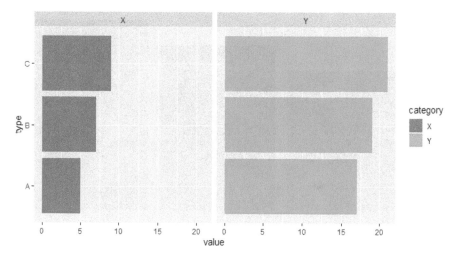

Fig. 1.3 Faceted horizontal column chart

Compare two plots side by side (Fig. 1.3).

```
# plot straight from data_unique
data_unique %>%
    # set key aesthetics (axes and color groups)
    ggplot(aes(x = type, y = value, fill = category)) +
    # create separate plots
    facet_wrap(vars(category)) +
    # chose chart type, and specify stacked
    geom_col(position = "stack") +
    # make bars horizontal
    coord_flip()
```

1.9.6 Using Purrr and Map for Iteration

The *purrr* package's *map()* family of functions are powerful, consistent, and intuitive for iterating functions and operations. There exist many advanced applications of the *purrr* package, and some introductions to the package online may seem daunting. We'll provide some simple examples here to illustrate how we'll utilize this package throughout the book. Once you get the hang of it, further exploring materials available online would be a good idea.[5]

Use *map()* to create a list of averages.

```
# create a few vectors
```

---

[5] purrr tutorial. Lessons and Examples. https://jennybc.github.io/purrr-tutorial/index.html

```r
a <- c(1,2,3)
b <- c(2,3,4)
c <- c(3,4,5)
# place them in a list
l <- list(a,b,c)
# map() will perform the same function on each element of the list.
# find the mean value of the elements in each vector,
# and notice that map places all the results into a list
map(l, mean)
# now place all the results into a numeric vector
map_dbl(l, mean)
# map() will also perform the same function on each element of
a vector
# map() is also smart enough to know that it should iterate only
# for the first argument of the function you specify
v <- 1:10
# repeat each element of the vector 3 times
map(v, rep, 3)
```

Create your own, more complicated function, and iterate that.

```r
# create a function that divides the 'value' column
# of the dataset we created above, by a specified number
calc_fraction <- function(div_by, df) {

  new_data <- df %>%
    mutate(fraction = glue("1 / {div_by}"),
           frac_value = value/div_by)

  return(new_data)
}
# create a vector of numbers to divide by
div <- c(1, 2, 3)
# perform the operation on data_unique for each value
# listed in the div vector. map() will store each resulting
# data frame as an element in a list
map(div, calc_fraction, data_unique)
# now stack the three results, row-wise in a new data fram
map_dfr(div, calc_fraction, data_unique)
```

The beauty of *map()* is its simplicity and speed for repeatedly performing some-times quite complex tasks and operations. No matter how long the vector *div* is above—i.e., the number of times you need to iterate—the code is the same. So any time you find yourself needing to repeat or iterate some task or function, the *map()* family of functions provides a simple and intuitive solution. For further tutorials on R applications, please visit ikenresearch.com/tutorials.

1.10   Reflections on the Tidyverse in Practice

The Tidyverse collection of packages is evolving, which is a risk to the reproducibility of analyses. New functions continue to emerge, and older ones are retired. For that reason, it's good to know that these scripts were created under R versions 4.0.2 and Tidyverse version 1.3.0.

Maintainers of the Tidyverse, though, are usually very good about building in warning messages about depreciated functions and how to correct for any errors that result from them. Depreciated functions are further only discontinued after long periods of time. The risk of encountering an intractable problem because of new package versions is thus very low. The benefit you gain from having more legible code—especially when learning new techniques—significantly outweighs the risk.

Inputs and Outputs

2

2.1 Objective

This chapter introduces the input-output paradigm and demonstrates key benefits of framing institutions in terms of inputs and outputs for performance assessment. By the end of the chapter, the reader will have a concrete structure for managing the complex organizational data needed for conducting a productivity analysis. The input-output paradigm serves as the foundation for assessing productivity. This chapter highlights core strengths and limitations of the input-output frame and guides the reader through the initial procedural steps of conducting a quantitative productivity analysis for an NFP or education institution. The chapter concedes that inputs and outputs alone cannot paint a complete picture of performance, quality, or outcomes, but the input-output perspective establishes reference points for conversation and benchmarking. The input-output perspective opens the door for measurement and objectivity.

2.2 What Are Inputs and Outputs?

Inputs represent the resources needed to perform an organization's key work processes. Outputs represent the direct products or services that result from the key work processes. Inputs and outputs are best understood when you can draw lines in the sand immediately before work commences and right when it ends. They specify what is needed for performing a task and what can be shown for having completed it.

Inputs may include human resources, capital, technology, finances, or intellectual property. Outputs may include services delivered, products created, objectives achieved, or targets met.

Determining which organizational elements are inputs and which are outputs, however, may not always be so straightforward. Different perspectives on issues or

© The Author(s), under exclusive license to Springer Nature
Switzerland AG 2021
K. Moore, *Measuring Productivity in Education and Not-for-Profits*,
Management for Professionals, https://doi.org/10.1007/978-3-030-72965-3_2

intermediate tasks can muddy the analytical waters. In NFPs and education institutions, work does not happen in perfect temporal fashion like on an assembly line. For example, delivering a workshop may be best understood as either an input or as an output depending on programmatic objectives. If a workshop concludes with some type of certification, and awarding certifications is the prime objective of a program, then workshop delivery might best be understood as a programmatic output. On the other hand, if a workshop is designed to build some foundational skill to enable subsequent action or to change behavior, then the workshop might best be understood as an input.

Organizational and programmatic context will always serve as the best guide for labelling elements of key work processes as inputs or outputs. Taking context into account, the input-output perspective serves no more important role than to catalyze rigorous thinking around complex organizational tasks. If poorly implemented, an input-output framing may be critiqued as too simple, but it is the simplicity of the paradigm that allows for useful unpacking of complexity.

2.3   A Detailed Look at Inputs and Outputs

2.3.1   The Input-Output Paradigm

To frame an organization in terms of its inputs and outputs represents a deliberate analytical choice with benefits and limitations. The prime benefit is that the input-output frame supports empirical and objective analysis. Without defining inputs and outputs, organizational productivity and performance can remain nebulous concepts, amenable to debate. Input-output framings provide anchor points for more directed conversations around performance.

A key limitation is that many qualities of inputs and outputs are difficult to define or measure. Such is often the case with NFPs and education institutions. Productivity analyses should be supplemented with additional qualitative insights to paint more comprehensive pictures of performance. The more difficult it is to measure inputs and outputs, the more important it is to gather additional forms of evidence.

Framing an organization in terms of its inputs and outputs allows for a rigorous thought exercise, although for an NFP or education institution with more human-centric aims and activities, the process may seem too mechanical or too blunt. The end goal of measuring productivity and performance, however—at least in this book—is always to improve productivity and performance. If we cannot accurately gauge our current state, then we have no frame of reference for evidencing whether or not it has improved after some period of time, in relation to peers, or as a result of some intervention or initiative.

To build a shared understanding of what this book means by a productivity improvement, it is important to understand that the input-output paradigm espoused here places equal importance on both the quality and quantity of inputs and outputs. Within the framing of this book, it is thus nonsensical to state that a productivity improvement could be a bad thing. Some colloquial uses of the term "productivity" may paint it in a negative light, insinuating that productivity improvements may lead to deteriorations in quality or a "race to the bottom." This book, however, does not view cutting costs at the expense of quality as a productivity improvement. Such an intervention may amount to "efficiency" gains, but not an improvement to productivity. The distinction between efficiency and productivity in this book is crucial as we apply the concept of productivity to NFPs and education institutions. Productive processes—as opposed to efficient processes—must have quality baked in.

2.3.2 Identifying Inputs and Outputs

A key characteristic of inputs and outputs is that they accumulate. Not any quantitative indicator is suitable to represent an organizational input or output. Ratios, rates, and rankings are never organizational inputs or outputs. They may serve as perfectly reasonable performance indicators for other phenomena, but inputs and outputs represent physical things or isolated achievements that you can count with your fingers. Dollars and cents, numbers of staff, degree completions, and research publications may all be considered inputs or outputs. Student-to-staff ratios, completion rates, and global rankings while informative are not inputs or outputs. As we will discuss in more detail in the next chapter, a productivity indicator itself is a ratio of outputs to inputs, but the input and output components themselves should not be ratios.

A search through the literature on productivity may reveal studies that have used student-to-staff ratios as inputs to a productivity analysis, as a representation of school quality. Such practice exemplifies an inherent pitfall of translating real-world phenomena into the language of mathematics. Once you start turning actual things into numbers, you may then add, subtract, multiply, or divide those numbers to your heart's content. But to produce an analysis with robust, reliable, and useful findings, quantitative characterizations of real-world phenomena must operate within bounds.

Lines will inevitably blur, however, with complex organizations and different objectives for analysis. With complex organizations, such as universities, there exist both disparate and interconnected streams of academic work with intermediate inputs and outputs. For example, the output of a community engagement program might also be considered an input to a student recruitment strategy. The awarding of competitive research funds might be considered the output of a lengthy grant application process, but those funds are also an input to subsequent research activities. Further, a commencing student may be considered an input to the learning process, but a total count of the student body serves as an output proxy for the scale of educational services being delivered. Analytical objectives and the type of productivity being measured must be articulated before inputs and outputs are selected.

2.4   Tutorial: Organizing Input-Output Data

2.4.1   Setup

This tutorial requires activating the packages listed below. If you do not have these packages installed, please do so using the `install.packages()` function.

```
library(tidyverse)
library(knitr)
library(janitor) # handy for tidying data. We'll use the clean_
names() function.
library(ggplot2) # visualizing data
```

2.4.2   Introduction

Our first tutorial begins with indicative data from two school districts of similar size. The scenario is such that state officials in the USA wish to benchmark performance at the district level within their state. They want to better understand the level of effort and resources required to generate expected student outcomes within different geographical locations and education contexts. Their aim is to highlight areas of high performance and opportunities for improvement. They need multiple views and perspectives on productivity. We thus need to organize multiple data sets to select appropriate inputs and outputs for examination. This means looking at finances, student and staff numbers, and learning outcomes. This chapter will focus first on the finances and student and staff numbers.

A trademark of any large institution or system is that data on performance often comes from multiple disparate sources. Some of the sources may be spreadsheets from department managers, exports from internal enterprise software, or downloads from public repositories or websites. This means that data will be coming at you in all forms, shapes, and sizes, and you need to be able to distill key inputs and outputs from a mash-up of different sources.

We begin the tutorial by creating a few mock data sets indicative of what they may look like once you import data into R. As covered in the previous chapter, we assume a basic knowledge of R functionality in this tutorial, such as importing data, so we move straight to the data wrangling.

We use a typical income and expenditures report structure as an example for how to filter and reshape key input and output variables for impending productivity analysis. First, most finance reports are published with the intent to be human readable. They are often organized in Excel so that an executive or elected official can scan through them and quickly get important pieces of information. They often have rows with bolded text to highlight broader expenditure or revenue categories. And often clear variable names for each column are not provided. Rather, columns of financial information are often associated with a point in time. This makes year-on-year comparisons easy, but data shaped in this way is not conducive to rigorous and efficient analysis. So let's see what this may look like in within the R environment.

We use the tribble() function to create a data set in a way that's visually easy to read.

```
# mock data
finance_data_raw <- tribble(
  ~x1,              ~x2,                ~x3,                    ~`2019`,
  ~`2020`,
  "District A",   NA,               NA,                    NA,      NA,
  NA,           "Total Expenditure", NA,                        35000000,
  36000000,
  NA,                    NA,              "Teaching salaries",
  12000000,  12500000,
  NA,                    NA,              "Non-teaching salaries",
  8000000,  8000000,
  NA,                    NA,              "Operating expenses",
  8000000,  8500000,
  NA,                    NA,                 "Depreciation",
  4000000,  4000000,
  NA,                    NA,                "Finance expense",
  2000000,  2000000,
  NA,                    NA,                  "Other costs",
  1000000,  1000000,
  NA,           "Total Revenue",      NA,                  40000000,
  41000000,
  NA,                    NA,              "Government revenue",
  38000000,  38000000,
  NA,                    NA,                 "Other income",
  2000000,  3000000,
  "District B",   NA,               NA,                    NA,      NA,
  NA,           "Total Expenditure", NA,                        43000000,
  44000000,
  NA,                    NA,              "Teaching salaries",
  15000000,  16000000,
  NA,                    NA,              "Non-teaching salaries",
  10000000,  10000000,
  NA,                    NA,              "Operating expenses",
  9000000,  9000000,
  NA,                    NA,                 "Depreciation",
  3000000,  3000000,
  NA,                    NA,                "Finance expense",
  4000000,  4000000,
```

```
  NA,                    NA,                    "Other costs",
2000000,    2000000,
  NA,             "Total Revenue",    NA,                40000000,
40000000,
  NA,                    NA,              "Government revenue",
39500000,   39500000,
  NA,                    NA,                  "Other income",
500000,     500000
)
```

2.4.3 Clean the Data

We need strategy for getting this data into better form. There are a few key problems to overcome. First, the data are not "tidy." That is, the data frame is not organized such that each column is a distinct variable and each row is a distinct observation. The second problem is that some of the observations are sums of the other observations. While this may be helpful for an executive or an auditor, it is not helpful for an analyst. While it may seem cumbersome, appropriately reshaping your data sets up the entire analytical project for success. Analysts often joke, 90% of data science is cleaning data, and the other 10% of data science is complaining about cleaning data.

First, we need a strategy for wrangling this data set. Notice that the first column is associated with the overarching type of financial information, revenue, and expenditure. The second column names a broad category, and the third column gives the most granular breakdown. So the first thing we need to do is rename these columns. Second, notice all the NA values. These often occur when someone has merged many different cells in an Excel spreadsheet. Thus what we need to do is populate all the NA values with the appropriate corresponding information. The fill() function from the tidyverse package allows us to do that. Third, we then need to omit redundant information. Specifically, we need to filter out any rows with summed totals. If we feel the need to sum any data in future, we can always do that. Lastly, we need to create a new variable, called "year," so we can efficiently bundle up all the financial information into a single column. We do this using the pivot_longer() function. Finally, the head() function shows the top six rows of the resulting data frame, so we can see the changes, and the kable() function displays results in a neatly organized table (Table 2.1).

```
# name variable columns
colnames(finance_data_raw)[1:3] <- c("district", "type", "variable")

# populate NAs, filter out summed rows, create year variable
finance_data_clean <- finance_data_raw %>%
  fill(district, type) %>%
  mutate(type = str_remove(type, "Total ")) %>%
```

```
filter(!is.na(variable)) %>%
  pivot_longer(`2019`:`2020`, names_to = "year", values_to =
"value")
```

```
# check out the top few rows of the data set to see the result
head(finance_data_clean)
```

These few simple lines transform our data set into something workable. Every row is now a single observation associated with one unique piece of financial information. Every piece of financial information has four different categorical variables as descriptors, and no information is redundant or repeated. This is commonly referred to as "long format" data. For many computational purposes, this is the most efficient and direct way to structure data, including visualizing data with ggplot. In this format, reproducing aggregated totals again is also easy, in case we are interested in broader descriptions of the data. For example, let's reproduce those overarching revenue and expenditure totals with a few simple lines. We accomplish this by using the group_by() function to denote the variables we are interested in aggregating and then the summarise() function to compute the totals (Table 2.2).

```
# view an aggregated version of the data
finance_data_clean %>%
        group_by(district, year, type) %>%
        summarise(total = sum(value)) %>%
        ungroup()
```

Table 2.1   Top six rows of cleaned school district data

District	Type	Variable	Year	Value
District A	Expenditure	Teaching salaries	2019	12000000
District A	Expenditure	Teaching salaries	2020	12500000
District A	Expenditure	Non-teaching salaries	2019	8000000
District A	Expenditure	Non-teaching salaries	2020	8000000
District A	Expenditure	Operating expenses	2019	8000000
District A	Expenditure	Operating expenses	2020	8500000

Table 2.2   Revenue and expenditure totals

District	Year	Type	Total
District A	2019	Expenditure	3.5e+07
District A	2019	Revenue	4.0e+07
District A	2020	Expenditure	3.6e+07
District A	2020	Revenue	4.1e+07
District B	2019	Expenditure	4.3e+07
District B	2019	Revenue	4.0e+07
District B	2020	Expenditure	4.4e+07
District B	2020	Revenue	4.0e+07

2.4.4   Input-Output Structure

From here we need to be comfortable with input-output data in both long format, like we just created, and a standard wide format that will come in handy later for ease in making productivity calculations. With the long format, we only need to add one more categorical column that specifies whether each variable is an input or an output. We use "x" to denote an input and "y" to denote an output. We use the case_when() function to make the new categories explicit (Table 2.3).

```
# add a column to specify inputs and outputs
xy_finance_long <- finance_data_clean %>%
  mutate(nature = case_when(
    type == "Expenditure"         ~ "x",
    type == "Revenue"             ~ "y")) %>%
  select(district, year, nature, everything())

# view the result
head(xy_finance_long)
```

You should think of this structure as the master template for input-output data. Using this generic structure, it will be easy to add more data as it comes in and we will do so below. It'll thus be helpful to look at this data in a different way. Use the glimpse() function to get a bit more information about the data than using head(). As we make intermittent changes to the data, glimpse is a quick way to see changes and makes sure we have gotten the intended results and sometimes to troubleshoot. This new function places more of a focus on the variables we have created and some of their attributes, rather than giving just a preview of the top few lines of data. We'll shift to using this new function more commonly than head() and kable().

```
glimpse(xy_finance_long)
## Rows: 32
## Columns: 6
## $ district <chr> "District A", "District A", "District A",
## "District A", "D...
## $ year     <chr> "2019", "2020", "2019", "2020", "2019", "2020",
## "2019", "2...
```

Table 2.3 Input-output designation for data elements

District	Year	Nature	Type	Variable	Value
District A	2019	x	Expenditure	Teaching salaries	12000000
District A	2020	x	Expenditure	Teaching salaries	12500000
District A	2019	x	Expenditure	Non-teaching salaries	8000000
District A	2020	x	Expenditure	Non-teaching salaries	8000000
District A	2019	x	Expenditure	Operating expenses	8000000
District A	2020	x	Expenditure	Operating expenses	8500000

```
## $ nature    <chr> "x", "x", "x", "x", "x", "x", "x", "x", "x",
"x", "x", "x"...
## $ type      <chr> "Expenditure", "Expenditure", "Expenditure",
"Expenditure"...
## $ variable <chr> "Teaching salaries", "Teaching salaries",
"Non-teaching sa...
## $ value    <dbl> 12000000, 12500000, 8000000, 8000000, 8000000,
8500000, 40...
```

First notice that that data have been transposed to focus on all the variables. We can quickly see the dimensions of the data set and the type of each variable. The first five variables in the data are "character" columns or text columns, and the final variable is a "double" column or numeric. In the R language, double refers to "double precision," as opposed to an integer variable, which can only be whole numbers.

We now pivot the data into a standard input-output wide format that makes certain productivity calculations easier and more intuitive. Productivity analysis often calls for examination of more than one institution (or in the current case, school districts). It can be helpful then to see the data shaped in a way where one observation corresponds to one organization, and we can easily see all the inputs and outputs for that organization. What we need to do is spread the numeric data across multiple columns, where each column represents one of the input-output variables. To do so, we use the pivot_wider(). First, though, we make a few manipulations to the columns for better readability.

```
# specify inputs and outputs within the variable names
# by pasting together the values from the 'nature' and 'variable'
columns.
# Then pivot to wide format
xy_finance_wide <- xy_finance_long %>%
  mutate(variable = paste(nature, variable, sep = "_")) %>%
  select(-c(nature, type)) %>%
  pivot_wider(names_from = "variable", values_from = "value") %>%
  clean_names()

glimpse(xy_finance_wide)
## Rows: 4
## Columns: 10
## $ district                 <chr> "District A", "District A",
"District B", "...
## $ year                     <chr> "2019", "2020", "2019", "2020"
## $ x_teaching_salaries      <dbl> 12000000, 12500000, 15000000,
16000000
## $ x_non_teaching_salaries <dbl> 8e+06, 8e+06, 1e+07, 1e+07
## $ x_operating_expenses     <dbl> 8000000, 8500000, 9000000, 9000000
## $ x_depreciation           <dbl> 4e+06, 4e+06, 3e+06, 3e+06
```

```
## $ x_finance_expense          <dbl> 2e+06, 2e+06, 4e+06, 4e+06
## $ x_other_costs              <dbl> 1e+06, 1e+06, 2e+06, 2e+06
## $ y_government_revenue       <dbl> 38000000, 38000000, 39500000,
39500000
## $ y_other_income            <dbl> 2e+06, 3e+06, 5e+05, 5e+05
```

2.4.5 Add in More Data

A key premise of this book is that productivity for NFPs and education institutions cannot be established only through financial indicators. We need to supplement and sometimes substitute financial information with other types of material performance indicators. Thus, for our school districts, we add some student and staff data; that is, those who consume and deliver higher education services, respectively.

Above we used the tribble() function to create mock data for readability. Here, we use the tribble() function for the sake of efficiency. We're going to build a data set that's already in tidy long format to show how easy it is to combine data when you've organized it in a consistent way.

```
# mock data
xy_staff_student_long <- tibble(
  district    = c(rep("District A", 8), rep("District B", 8)),
  year        = c(rep("2019", 4), rep("2020", 4), rep("2019", 4),
rep("2020", 4)),
  nature      = c(rep(c("x","x","y","y"), 4)),
  type          = c(rep(c("Staff","Staff","Students","Stude
nts"), 4)),
    variable        = c(rep(c("Teaching","Non-teaching","Low
disadvantage",
                    "High disadvantage"), 4)),
  value       = c(300, 100, 6500, 2500, 305, 100, 6505, 2505,
                  325, 125, 4000, 5000, 325, 120, 4005, 5005)
)
```

```
# check out result
glimpse(xy_staff_student_long)
## Rows: 16
## Columns: 6
## $ district <chr> "District A", "District A", "District A",
"District A", "D...
## $ year     <chr> "2019", "2019", "2019", "2019", "2020", "2020",
"2020", "2...
## $ nature   <chr> "x", "x", "y", "y", "x", "x", "y", "y", "x",
"x", "y", "y"...
```

```
## $ type      <chr> "Staff", "Staff", "Students", "Students", "Staff",
"Staff"...
## $ variable <chr> "Teaching", "Non-teaching", "Low disadvan-
tage", "High disa...
## $ value      <dbl> 300, 100, 6500, 2500, 305, 100, 6505, 2505,
325, 125, 4000...
```

For staff data, we have FTE totals for teaching and non-teaching staff. For students we have indicators for the number of students with both low and high socioeconomic disadvantage. We now put all our data into one place, so we have the option to easily explore multiple input-output combinations. We append the staff and student data to the financial data using the bind_rows() function. After doing so, you will notice that the complete data set has grown to 48 rows.

```
# append the new data to our inital data set and sort
xy_data_long <- xy_finance_long %>%
  bind_rows(xy_staff_student_long) %>%
  arrange(district, year, nature, type, variable)


# check results
glimpse(xy_data_long)
## Rows: 48
## Columns: 6
## $ district <chr> "District A", "District A", "District A",
"District A", "D...
## $ year      <chr> "2019", "2019", "2019", "2019", "2019", "2019",
"2019", "2...
## $ nature    <chr> "x", "x", "x", "x", "x", "x", "x", "x", "y",
"y", "y", "y"...
## $ type      <chr> "Expenditure", "Expenditure", "Expenditure",
"Expenditure"...
## $ variable <chr> "Depreciation", "Finance expense", "Non-
teaching salaries"...
## $ value      <dbl> 4000000, 2000000, 8000000, 8000000, 1000000,
12000000, 100...
```

At this point, in your current working directory, create a new folder called "data." Then run the code below. If you run the code below before creating the "data" folder, you will get an error. We want to save is data set because it will come in handy in the next chapter.

2.4.6   Fashion a Data Set for Specific Analysis

Now that we have all our data in place, imagine a simple scenario where state officials want to benchmark performance in terms of expenditure. We can quickly filter the data to make some useful comparisons. We can then put ggplot2 to use to visualize for some quick interpretation (Fig. 2.1).

```
# filter for the variables of interest
chart_data <- xy_data_long %>%
  filter(type %in% c("Expenditure"))

# specify data, set the aesthetics (chart parameters),
# specify the chart type (column chart), and
# make horizontal
ggplot(data = chart_data,
       aes(y = value, x = variable, fill = district)) +
  geom_col(position = "dodge") + # code for grouped column chart
  coord_flip() # horizontal layout
```

After examining the differences in expenditure between districts, then appreciate how simple it was to create a chart from our long format data. The ggplot2 package is optimized for long format data, making it simple to specify all your chart parameters based on the categorical variables. Fashioning a compelling visual does take some getting used to, but the four lines of code above should illustrate how little effort can go into an effective chart, as long as your data are kept tidy. We'll learn some additional ggplot2 features and explore different visualizations as we go along.

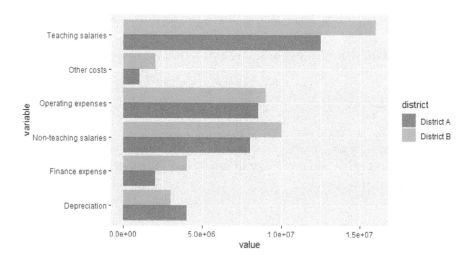

Fig. 2.1 School districts performance indicators comparison

2.5   Reflections on Inputs and Outputs in Practice

Inputs and outputs are useful but not perfect. Especially at this stage in the book, our examples are simple for demonstration purposes. The number of students at a school, for example, is not sufficient on its own to perfectly represent the volume of teaching delivery at a school. Assumptions about standard teaching hours per student, student-to-teacher ratios, and other factors would have to be considered for more valid and reliable findings. Broad comparisons of expenditure per student may, however, provide an interesting and appropriate place to begin exploring the data. No final conclusions should be reached from such a blunt analysis, but if findings reveal striking differences or unexpected gaps, then we now have an excellent anchor point for further conversation and deeper analysis. If the broad analysis reveals nothing exciting, then we may choose to focus on other variables of interest.

Input-output analysis is simple in concept but lends itself to generating deep and nuanced insight. At this stage, it may seem that we are only rebranding a simple and common metric, such as expenditure per student, into a more technical-sounding "input-output analysis." And at only this stage you would be correct! However, it takes some simple examples at the front end to lead into more compelling work. The input-output frameworks we are setting up will allow us to eventually explore multiple phenomena in conjunction with one another, weight their importance accordingly, and generate a suite of more tailored performance indicators. Choosing our inputs and outputs wisely and thus allowing us to examine productivity from multiple angles will lead to analyses that may explicitly show the successes and trade-offs that manifest when implementing different institutional or system-level strategies. First organizing our data into inputs and outputs provides a simple foundation for unpacking complexity.

An analyst or researcher wishing to proceed beyond this stage of analysis with a real-world organization should do so either with the support of organizational leaders or with the intention to garner that support. Findings from productivity analyses often lend themselves to difficult and objective conversations about performance and leverage points for change. The input-output perspective helps to frame organization problems and thereby helps to frame solutions. It is thus crucial (A) to ensure that multiple angles are covered with the input-output data under examination, (B) to acknowledge known limitations, and (C) to ensure an appropriate audience and context for sharing analytical results. Productivity portrayals can be hard-edged and immediately turn off stakeholders who are interested in less-measurable phenomena. But if you are part of an organization that cannot seem to consistently or objectively frame performance issues, and if you can present findings fairly, then strict input-output frameworks can lead to findings that will not allow stakeholders to turn their heads. On the other hand, if not handled with care, introducing hard quantitative indicators in environments where less tangible performance considerations are traditionally held in higher regard, then you may find significant pushback and difficulty in making a case and changing minds.

The Productivity Ratio

3

3.1 Objective

This chapter demonstrates the necessity of tackling education productivity from multiple angles. By the end of the chapter, the reader will understand how to address three competing questions about institutional performance using three bespoke measures of productivity. This chapter builds on the scenario from the previous chapter regarding two different, similarly sized school districts. It dives deeper into institutional data and uses productivity indicators (A) to benchmark institutional operations and (B) to approximate the effectiveness of learning and teaching. The benchmarks and approximations are achieved through creating several contrasting input-output ratios. Limitations of each method are discussed as well.

3.2 What Is the Productivity Ratio?

The productivity ratio is a measure of output to input, or O/I. It represents output produced per unit input consumed. If you hold inputs constant and increase output, then productivity is understood to increase. If you hold outputs constant, then inversely, you can increase productivity by reducing inputs.

Underpinning the productivity ratio is the idea of the "production function," or $f(I) = O$. The production function implies that outputs are a direct function of inputs and an institution's production technology. "Technology" in this sense is a broad term encompassing any range of potential processes or procedures—human or machine—that transform inputs into outputs.

The term "production technology" is less than ideal as applied to NFPs and education institutions, and we'll shy away from that language in this book. The benefit of thinking in these terms, however, is to set the tone for a technical, end-to-end analysis of the mechanisms and practices institutions put into place to achieve their

© The Author(s), under exclusive license to Springer Nature
Switzerland AG 2021
K. Moore, *Measuring Productivity in Education and Not-for-Profits*,
Management for Professionals, https://doi.org/10.1007/978-3-030-72965-3_3

aims. The technical definition is intended to draw focus away from political and interpersonal concerns and instead shift focus toward what has been achieved with the available resources. Once we have formed an objective response to that concern, we then have full license to drill down into the equally important political, interpersonal, or contextual issues that may have driven the result.

3.3   A Detailed Look at the Productivity Ratio

3.3.1   Total-Factor, Multi-Factor, and Single-Factor Productivity

Understanding the taxonomy around different types of productivity helps to open the door for different types of indicators and analysis. Total-factor productivity (TFP) is a measure that attempts to capture the full extent of inputs used to produce outputs. TFP is intended to account for all capital, labor, energy, materials, and services (KLEMS) employed during the production process. The most common proxy for KLEMS is the comprehensive statement of expenditures associated with output production. Multi-factor productivity (MFP) may be distinguished from TFP because, while MFP attempts to capture multiple inputs to the production process, it does not necessarily intend to be comprehensive. For example, there may be cases where we want to examine labor and materials, but we do not need to account for capital expenditure. Single-factor productive (SFP) accounts for only one type of input to the production process. The most common factor for SFP analysis is labor. Labor often represents the largest single item for organizational expenditure and is often most consequential input to production. It is thus common to examine separately. Recognizing that we have multiple options for characterizing productivity, it is frequently the case that certain contexts call for one measure over the other.

3.3.2   Selecting Appropriate Indicators for Education

Constructing a productivity ratio for an education institution or NFP requires attention to the specific question being asked about performance. Unlike a manufacturing plant producing a target number of widgets within a given timeframe—tested under strict quality standards—for education institutions not all stakeholders would even agree on what outputs are most important to track, and quality is always subject to debate and interpretation.

Even if we have identified certain indicators as inputs or outputs, using them appropriately for a productivity analysis requires a deeper level of consideration. Let's consider a few caveats for why input-output selection is so important for arriving at valid and compelling results. First, consider the student to teacher ratio as an indicator of labor productivity. For some schools or school districts, models of instruction, policies, or regulations may place strict floors or ceilings on the student to teacher ratio as standard for quality. In this case, it is not reasonable to use the student to teacher ratio as a valid productivity indicator. Not only may the school not have full agency to significantly change their student to teacher ratio, but on principle it is not necessarily reasonable to assume that a more productive school is always one with a higher value for this figure.

There also exist input variables with inherently complex relationships with outputs. Consider, for example, the number of core curriculum elements. A school's curriculum is an input to the learning process. And one can imagine that packing the curriculum with too much content and material would be counterproductive. Just as with a normal productivity ratio, increasing inputs fundamentally decreases the value of O/I. However, if you started reducing your curriculum to the point where teachers no longer have any set material to cover with their students, then this would also likely be counterproductive to the learning process. The sweet spot lies somewhere in how teachers choose to scaffold lessons and deliver material. This simply cannot be captured with a bidirectional indicator.

3.3.3 Value-Add

With respect to outputs, we are not always concerned about the end result in isolation. We often need to contextualize learning with respect to initial conditions and prior student knowledge. Understanding prior knowledge allows us to estimate the extent to which the learning process has improved students' capabilities. This concern relates to the concept of "value-add." For the productivity ratio, capturing value-add means adjusting the formula to something like $(O - I)/I$. An example from the private sector will help illustrate the difference.

In the for-profit world, value-add productivity may be as simple as tracking profit, instead of revenue, as the output. Consider a manufacturing company wanting to measure the productivity of its widget production process. The company first defines output as the total value of all widgets they produce and the input as the total cost of producing them. If the company produces 100 widgets to be sold at 10 USD apiece, and all production costs amount to 1000 USD, then we could define the productivity ratio as $O/I = 100 * 10/1000 = 1$. The productivity is thus one dollar earned for every dollar spent—assuming all units sell. However, if the company only breaks even, then have they succeeded in adding any value? A more appropriate productivity ratio might be $(O - I)/I = (100 * 10 - 1000)/1000 = 0$. The value-add productivity indicator is less complementary, but depending on targets and objectives, it may serve as a more appropriate judge of performance.

For education and students, understanding value-add means trying to understand how much a student has learned, rather than how much they know.

3.4   Tutorial: Competing Production Functions

3.4.1   Setup

This tutorial requires activating the packages listed below. It also requires the dataset we created in the previous chapter, xy_data_long. If you have not yet run the code from the previous chapter, you will need to do so before proceeding with this tutorial.

```
library(tidyverse)
library(knitr)
library(janitor)

load("data/xy_data_long.rda")
```

3.4.2   Introduction

In this tutorial we examine four separate representations of school productivity. We demonstrate how different cuts of data tell different stories about performance. Each productivity portrayal provides an alternative perspective and highlights the importance of addressing opposing interests of diverse stakeholders and decision-makers.

We build from the example in the previous chapter involving two similarly sized school districts in the USA. Assume that district administrators have the autonomy to utilize and allocate funds as they see fit across the districts' schools. Administrators in each district want to learn from the practices of the other to consider alternatives for utilizing resources. Their objective is to improve processes and outcomes in their schools.

Between the two districts, there are several outspoken personalities with influence in their communities. Some are concerned about efficiency and taxpayer money and want to ensure the schools are not running operational deficits. Some are more concerned about the districts' reputations and statewide rankings. Others are more concerned about equality of opportunity and supporting disadvantaged children in their communities as best as possible. It's our job to examine productivity and performance from each of these angles.

A final note on this scenario is that we have the results of a statewide pilot study to measure student learning during year-nine. Prior educational research suggested that success during year-nine is a strong predictor of continued educational success. The pilot project in the two districts thus sought to measure student learning during year-nine by introducing a pre-test at the beginning of the year to be compared to the results of a statewide standardized exam that already occurs at the end of year-nine. The new exam will allow for a pre-test, post-test comparison.

3.4.3   Create and Join the New Test Scores Data

Our first step is to create a mock data set of pre- and post-test results. We'll use the same data structure as in the previous chapter, so the new data is easy to join with our previous data, xy_data_long.

```r
xy_scores_long <- tibble(
  district     = c(rep("District A", 10), rep("District B", 10)),
  year         = c(rep("2020", 20)),
  nature       = c(rep("x", 5), rep("y", 5), rep("x", 5),
rep("y", 5)),
  type         = c(rep("Pre-test", 5),
                   rep("Final exam", 5),
                   rep("Pre-test", 5),
                   rep("Final exam", 5)),
  variable     = c(rep(c("1","2","3","4","5"), 4)),
  value        = c(500, 1000, 2000, 3500, 2000,
                   500, 1500, 2500, 2000, 2500,
                   2000, 3000, 2500, 1000, 500,
                   1500, 2500, 3000, 1000, 1000)
)
```

```r
glimpse(xy_scores_long)
## Rows: 20
## Columns: 6
## $ district <chr> "District A", "District A", "District A",
"District A", "D...
## $ year     <chr> "2020", "2020", "2020", "2020", "2020", "2020",
"2020", "2...
## $ nature   <chr> "x", "x", "x", "x", "x", "y", "y", "y", "y",
"y", "x", "x"...
## $ type     <chr> "Pre-test", "Pre-test", "Pre-test", "Pre-test",
"Pre-test"...
## $ variable <chr> "1", "2", "3", "4", "5", "1", "2", "3", "4",
"5", "1", "2"...
## $ value    <dbl> 500, 1000, 2000, 3500, 2000, 500, 1500, 2500,
2000, 2500, ...
```

First notice the new data comes from only the year 2020, when the pilot was conducted. The type column indicates whether scores are from the pre-test or final exam. The variable column lists the standardized scores that students could receive on the exams. The value column lists the total number of students from the districts who earned each listed score. Let's now join the new data to our full productivity dataset.

```
xy_data_long <- xy_data_long %>%
  bind_rows(xy_scores_long)
```

3.4.4   Operational Efficiency Productivity

Recall from our scenario above that there exist at least three distinct opposing interests in school performance. The first mentioned was an interest in operational efficiency. In order to assess this, we'll construct two productivity ratios, one to measure TFP and the other to estimate SFP for labor productivity. In both cases, the total number of students will serve as the output proxy for the scale of educational services delivered.

First prepare the data. We'll select the variables of interest, and use the technique from the previous chapter to pivot the data to wide format (Table 3.1).

```
# select key variables, summarise to get totals,
# and pivot to wide format
xy_ops_eff_wide <- xy_data_long %>%
  filter(type %in%  c("Students", "Staff", "Expenditure")) %>%
  group_by(district, year, nature, type) %>%
  summarise(value = sum(value)) %>%
  ungroup() %>%
  mutate(type = paste(nature, type, sep = "_")) %>%
  select(-c(nature)) %>%
  pivot_wider(names_from = "type", values_from = "value") %>%
  clean_names()

xy_ops_eff_wide
```

Now that we have a simpler set of data, we can create our benchmark productivity indicators (Table 3.2).

```
prod_ops_eff <- xy_ops_eff_wide %>%
  mutate(TFP_scale_service = y_students/x_expenditure,
         SFP_scale_service = y_students/x_staff) %>%
  select(district, year, TFP_scale_service, SFP_scale_service)

prod_ops_eff
```

Table 3.1 Input-output designation for operational efficiency

District	Year	x_expenditure	x_staff	y_students
District A	2019	3.5e+07	400	9000
District A	2020	3.6e+07	405	9010
District B	2019	4.3e+07	450	9000
District B	2020	4.4e+07	445	9010

Table 3.2 'Operational efficiency productivity' scores comparison

District	Year	TFP_scale_service	SFP_scale_service
District A	2019	0.0002571	22.50000
District A	2020	0.0002503	22.24691
District B	2019	0.0002093	20.00000
District B	2020	0.0002048	20.24719

Table 3.3 'Operational efficiency productivity' normalized

District	Year	TFP_norm	SFP_norm
District A	2019	1.0000000	1.0000000
District A	2020	0.9733025	0.9887517
District B	2019	0.8139535	0.8888889
District B	2020	0.7963384	0.8998752

How should we interpret and compare these results? Because the productivity indicators have been created using different units, the scales of results differ considerably. The actual numerical values of our new productivity indicators, however, are not necessarily important in themselves. The literal meaning of the TFP indicator, for example, is that for each dollar spent, you can educate about two ten-thousandths of a student—not particularly helpful.

Important to recognize is that our new productivity indicators have been created, such that they are directly comparable to one another in an absolute sense. For example, if the TFP value of District A is twice that of District B, it means that District A is twice as productive as District B along that particular dimension of performance.

To convert the TFP and SFP indicators into something more intuitive, we can normalize the scores across both schools and both years, with respect to the maximum value in each column. In data science, you'll often see adjustments like this referred to as feature scaling. Feature scaling, or normalizing, can be helpful under a variety of circumstances because it transforms the dataset to something more manageable while preserving all variation in the data and the relationships between elements. In effect, it accomplishes the same thing as just zooming out and getting a bird's eye view of the data (Table 3.3).

```
# divide the scores in each column by the max value for that column
prod_ops_eff_norm <- prod_ops_eff %>%
  mutate(TFP_norm = TFP_scale_service/max(TFP_scale_service),
         SFP_norm = SFP_scale_service/max(SFP_scale_service)) %>%
  select(district, year, TFP_norm, SFP_norm)

prod_ops_eff_norm
```

Table 3.4 Potential student factors influencing productivity

District	Year	High disadvantage	Low disadvantage
District A	2019	0.28	0.72
District A	2020	0.28	0.72
District B	2019	0.56	0.44
District B	2020	0.56	0.44

Now we can more easily compare the two districts. First, District A is shown to be more productive along both performance dimensions. Recall that TFP attempts to capture the effectiveness of total resource use in delivering educational services, whereas SFP attempts to capture the effectiveness of labor utilization.

A face value reading of these indicators goes as follows. For TFP the highest level of productivity was achieved by District A in 2019, and its productivity fell about 3% in 2020. District B was approximately 81% as efficient as District A in 2019, and its productivity fell less than 2% in 2020. For SFP, District A's score in 2019 again sets the benchmark, and its productivity fell by only 1% in 2020. District B was operating at about 90% productivity as compared to District A in both years.

These broad indicators should never be the end of the story, but we have created some hard indicators for effectiveness levels of resource use in both districts. We should now ask what factors may help explain these results? Do additional contextual factors not present in the data need to be investigated? One interesting driver might be the larger proportion of disadvantaged students attending District B. There could be extra staffing and resourcing requirements to accommodate for special needs. The code below helps to illustrate the difference. While District A has fewer than 30% of students with high disadvantage, more than half of District B's students have high disadvantage (Table 3.4).

```
# filter for only students, find the proportion of
# high and low disadvantaged students in each district and year,
# pivot to wide format
student_breakdown <- xy_data_long %>%
  filter(type == "Students") %>%
  group_by(district, year) %>%
  mutate(prop = round(value/sum(value), 2)) %>%
  select(-nature, -type, -value) %>%
  pivot_wider(names_from = "variable", values_from = "prop")

student_breakdown
```

3.4.5   Reputational Productivity

The second stakeholder interest mentioned in the scenario description above was an interest in district reputation and statewide rankings for student performance on standardized tests. Some stakeholders believe that teacher pay and student performance should be linked. They want to see more students performing in the top 4th and 5th quintiles (above the state average) on a standardized final exam, and they believe school human capital investments to be a prime mechanism from achieving this.

There exists a large amount of criticism regarding the use of standardized tests in the USA, especially regarding their use as performance management tools for schools. Despite criticisms and limitations, however, using test scores as a proxy for student learnings still represents an incremental step in the right direction from using only the productivity measures above to gauge performance, which included only student, staff, and financial quantities.

To gauge reputational performance, we will create a new type of labor productivity ratio, using only teaching salaries as inputs and the amount of top performing students the district produces as outputs. As crude as it may sound, we want to see what the numbers say when we ask, "how many top performing students do your staff salaries buy you?"

```
# filter for teacher salaries and students within
# the top 4th and 5th quintiles of the final exam,
# group those students as top students, and
# summarise results
xy_staff_top_long <- xy_data_long %>%
    filter (year == "2020",
            type %in% c("Expenditure", "Final exam"),
            variable %in% c("Teaching salaries", "4","5")) %>%
    mutate(variable = str_replace(variable, c("4", "5"), "Top
studs")) %>%
    group_by(district, year, nature, type, variable) %>%
    summarise(value = sum(value)) %>%
    ungroup()

glimpse(xy_staff_top_long)
## Rows: 4
## Columns: 6
## $ district <chr> "District A", "District A", "District B",
"District B"
## $ year     <chr> "2020", "2020", "2020", "2020"
## $ nature   <chr> "x", "y", "x", "y"
```

Table 3.5 Alternative input-output designation for school districts

District	Year	x_teaching_salaries	y_top_studs
District A	2020	12500000	4500
District B	2020	16000000	2000

```
## $ type       <chr> "Expenditure", "Final exam", "Expenditure",
"Final exam"
## $ variable <chr> "Teaching salaries", "Top studs", "Teaching
salaries", "To...
## $ value    <dbl> 12500000, 4500, 16000000, 2000
```

Now, put the data into wide format (Table 3.5).

```
# pivot to wide format
xy_staff_top_wide <- xy_staff_top_long %>%
  mutate(variable = paste(nature, variable, sep = "_")) %>%
  select(-c(nature, type)) %>%
  pivot_wider(names_from = "variable", values_from = "value") %>%
  clean_names()
```

```
xy_staff_top_wide
```

We can immediately see that District B is spending more on teacher salaries but generating fewer top students. Calculating and normalizing the productivity ratios will help us see the extent of the difference between the two (Table 3.6).

```
# calculate the ratios and normalize
prod_staff_top <- xy_staff_top_wide %>%
  mutate(SFP_top_studs = (y_top_studs)/x_teaching_salaries,
         SFP_top_norm = SFP_top_studs/max(SFP_top_studs)) %>%
  select(-x_teaching_salaries, -y_top_studs)
```

```
prod_staff_top
```

Incorporating learning outcomes has resulted in an even bleaker picture for District B's performance than the operational efficiency indicators from the first section of the tutorial. District B's reputational productivity is only about 35% of that of District A. If we stopped our analysis here, we might conclude that District B is inefficient and overpaying its staff. However, we know there are other factors at play. Namely, the two districts are operating within different contexts, as evidenced by the striking difference between the compositions of their student populations.

Table 3.6 Reputational productivity' comparison

District	Year	SFP_top_studs	SFP_top_norm
District A	2020	0.000360	1.0000000
District B	2020	0.000125	0.3472222

3.4.6 Value-Add Productivity

We now move to the interests of stakeholders more concerned with equity and with the true value-add of the educational processes occurring within the districts' schools. Using our available data, we will define a measure that accounts for district productivity with respect to changes in student performance between the year-nine pre-test and post-test. We assume for the sake of simplicity that all students have taken both tests, so there can be direct comparisons for all students regarding their scores.

To capture value-add, we need to estimate an average level of capability across all students according to the pre-test and an average level of capability according to the post-test. This will serve as the output (or numerator) of our productivity ratio. Remember, though, both outputs and inputs need to be measured in meaningful units that can accumulated, so the value-add indicator should be constructed in a way that tracks both the quantity of students and the extent to which they improved. For the input, we will again use teaching salaries, so we can directly juxtapose the indicator we create in this section with that of the previous section, measuring reputational productivity.

First, filter for the pre- and post-test data. Then account for only year-nine students. For simplicity, we assume that year-nine students represent one fourth of the student body in both districts.

```
# filter and adjust
xy_prepost_long <- xy_data_long %>%
  filter (year == "2020",
          type %in% c("Final exam", "Pre-test")) %>%
  mutate(value = value*(1/4))

# view result
glimpse(xy_prepost_long)
## Rows: 20
## Columns: 6
## $ district <chr> "District A", "District A", "District A",
"District A", "D...
## $ year     <chr> "2020", "2020", "2020", "2020", "2020", "2020",
"2020", "2...
## $ nature   <chr> "x", "x", "x", "x", "x", "y", "y", "y", "y",
"y", "x", "x"...
## $ type     <chr> "Pre-test", "Pre-test", "Pre-test", "Pre-test",
"Pre-test"...
## $ variable <chr> "1", "2", "3", "4", "5", "1", "2", "3", "4",
"5", "1", "2"...
## $ value    <dbl> 125, 250, 500, 875, 500, 125, 375, 625, 500,
625, 500, 750...
```

Now find the pre- and post-test results across the two districts. Do so by creating a new variable, called student_points, which is the product of the number of students and their respective scores. A higher score represents more educational value, and the more students who have achieve a higher score, the more value we assume there is. In the chunk of code below, we'll also find the average score per district just as a reference (Table 3.7).

```
# create the student_points variable,
# summarise to find the total number of student points, and
# calculate average scores per district.
xy_prepost_diff <- xy_prepost_long %>%
  mutate(student_points = as.numeric(variable)*value) %>%
  group_by(district, year, nature, type) %>%
  summarise(students = sum(value),
            student_points = sum(student_points)) %>%
  ungroup() %>%
  mutate(average_score = student_points/students)

xy_prepost_diff
```

We see now that, although students from district B have lower final exam scores on average, they have improved their standing relative to other districts in the state from pre-test to post-test. Conversely, while District A has higher scores on average, their state standing has actually fallen from pre-test to post-test. As stated above, the value-add will incorporate the difference between total students' points from pre-test to post-test, and we can infer from the indicators above that District A's value-add will be negative (Table 3.8).

```
# select key variables,
```

Table 3.7 Aggregate pre-test and post-test score comparison

District	Year	Nature	Type	Students	student_points	average_score
District A	2020	x	Pre-test	2250	8125	3.611111
District A	2020	y	Final exam	2250	7875	3.500000
District B	2020	x	Pre-test	2250	5500	2.444444
District B	2020	y	Final exam	2250	6125	2.722222

Table 3.8 Value-add variable designation

District	Year	y_value_add
District A	2020	-250
District B	2020	625

```r
# spread to wide format,
# and calculate the value-add output.
xy_prepost_wide <- xy_prepost_diff %>%
  mutate(type = paste(nature, type, sep = "_")) %>%
  select(-c(nature, students, average_score)) %>%
  spread(key = "type", value = "student_points") %>%
  clean_names() %>%
  mutate(y_value_add = y_final_exam - x_pre_test) %>%
  select(-c(y_final_exam, x_pre_test))

xy_prepost_wide
```

We now have our value-add outputs. What these figures mean is that on average in District A, 250 students dropped to one quintile lower from pre-test to post-test. In District B, however, on average 625 students moved one quintile higher from pre-test to post-test. We can now compare these outputs to our staff salaries input, so we can compare the differences in productivity (Table 3.9).

```r
# join in the input values from the section above, and
# calculate the new productivity scores
prod_value_add <- xy_staff_top_wide %>%
left_join(xy_prepost_wide) %>%
  mutate(SFP_value_add = y_value_add/x_teaching_salaries,
         SFP_value_norm = SFP_value_add/max(abs(SFP_value_add))) %>%
  select(district, year, SFP_value_add, SFP_value_norm)

prod_value_add
```

We can see now that from a yet another perspective on productivity, District A's performance portrayal has transformed from the overwhelming productivity leader to being counterproductive by a factor of about 0.5. In other words, when accounting for initial conditions and a proxy for student learning, for every step forward taken by District B, District A takes half a step backward.

3.4.7 Compile All the Results

Since we have data for both districts across all indicators in the year 2020, we will compare all our results side by side from that year. We use ggplot() to visualize, and we demonstrate how you can take a dataset and pipe straight to the chart (Fig. 3.1).

Table 3.9 'Value-add productivity' comparison

District	Year	SFP_value_add	SFP_value_norm
District A	2020	-2.00e-05	-0.512
District B	2020	3.91e-05	1.000

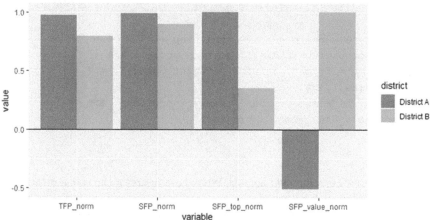

Fig. 3.1 Four different perspectives on Disctrict A and B's productivity

```
# join all results,
# select only normalised productivity scores,
# pivot to long format for plotting
all_results <- prod_ops_eff_norm %>%
  filter(year == 2020) %>%
  left_join(prod_staff_top) %>%
  left_join(prod_value_add) %>%
  select(-c(year, SFP_top_studs, SFP_value_add)) %>%
    pivot_longer(TFP_norm:SFP_value_norm, names_to = "variable",
values_to = "value")

# arrange the productivity scores in the order created, and
# pipe directly to ggplot to visualise
all_results %>%
  mutate(variable = factor(variable, levels = c("TFP_norm", "SFP_
norm", "SFP_top_norm", "SFP_value_norm"))) %>%
  ggplot(aes(x = variable, y = value, fill = district)) +
  geom_col(position = "dodge") +
  geom_hline(yintercept = 0) +
  labs(title = "Productivity meausures for District A and District
B in 2020")
```

3.5   Reflections on the Productivity Ratio in Practice

The examples here were designed to mimic circumstances that play out repeatedly in the real world. Each subsequent addition of new data can tell a different story. It is often not the job of the analyst to determine which performance portrayal is more. It is the job of the analyst to incorporate each new perspective, synthesize the variety of results, and attempt to paint a more complete picture of performance as new observations reveal new relationships.

In quantitative studies on student and school performance, controlling for student attributes is often considered an essential step in analysis, that is, accounting for different kinds of students based on their backgrounds or other qualities as they relate to measured results. Common patterns with different "types" of students are often factored in to an analysis in attempt to tease out the genuine effects of the learning process. It will always be important to try and understand the differential effects of any kind of intervention on different kinds of individuals. Schooling does not occur in a vacuum or a petri dish, so we must always consider, "what works, for whom, and under which circumstances?"

The analysis above shows how careful treatment of high-level information can also paint nuanced and opposing pictures of performance. Each portrayal of productivity in this chapter is valid in its own right and serves to address its own unique corresponding question. It is the marriage of all the factors here that would allow leaders and administrators to implement more informed decisions.

Multiple simultaneous production functions are always playing out in each school. There is a financial production function to be measured in earnest regarding expenditures and revenues, and there is another production function regarding the learning process. It may be tempting to want a single indicator of overall performance or productivity, but aggregating all the data in this way would likely hide important relationships. In a high-functioning school, the financial and student learning production functions may be mutually reinforcing, where good teaching leads to more students, leads to more revenue, leads to better resources, and leads to better teaching. In a different kind of school, these two functions may be at odds with one another, where differential pay between staff leads to contempt, leads to a lack of focus on learning and teaching, leads to poor student outcomes, and leads to more inequitable pay.

The moral of this story is that we need to be careful and intentional when we are mixing and matching different inputs and outputs. Schools are inherently complex, and even forming a hypothesis about what is going on in one can be difficult. Each new cut of data has the potential to tell a different part of the story. We should explore different kinds of indicators and relationships between variables, but analysts have a responsibility to not mislead decision-makers about some perceived importance or magnitude of a single measured relationship. It is the analyst's job to provide the anchor point for further conversation, acknowledge limitations of the data, and make clear where key quantitative information is missing, so that it can be gathered in a different way.

Productivity Change

4

4.1 Objective

The chapter characterizes university performance in terms of how productivity changes over time. By the end of the chapter, the reader will have become acquainted with an industry standard technique for measuring productivity change, the Törnqivst Index (TI). TI methods are espoused by the OECD and used by Bureaus of statistics across the globe to measure productivity trends in various sectors of the economy. This chapter covers the strengths and limitations of applying TIs to university data.

4.2 What Is Productivity Change?

Productivity change measurements indicate the difference between levels of productivity at two different points in time or across a period. Investigating productivity change means examining time series data and the shape of trends. The measurement exercise helps to reveal dynamic relationships between data elements and uncover the key factors governing observed behavior.

When measuring productivity change, the investigator is less concerned about making definitive statements about whether or not one organization is more productive than another. Rather, the objective is to determine which organizations are changing and at what rate. The benchmark is past performance, and individual organizations are generally held up to themselves, rather than to others. Institutions may be compared to one another, but rather than comparing absolute performance levels, the interest is shifted toward the relative steepness of curves. Institutions are portrayed only with respect to their rates of change.

Performing such an analysis can address two prime issues regarding NFP and educational productivity measurement. The first is that results of productivity

© The Author(s), under exclusive license to Springer Nature Switzerland AG 2021
K. Moore, *Measuring Productivity in Education and Not-for-Profits*, Management for Professionals, https://doi.org/10.1007/978-3-030-72965-3_4

change measurement can reveal a great deal about the interworkings and performance trajectories of organizations without the need to make claims about one organization's absolute performance over another. As we saw in the previous chapter, the numerous ways in which productivity can be measured, the different operational contexts, and the diverse interests of stakeholders can confound any analysis which seeks to determine which institution is more productive than another. NFPs and educational institutions cannot be judged with a one-size-fits-all approach, but leaders and stakeholders may still be interested in concrete measures of performance and productivity. Measuring productivity change allows for a level of precision akin to that of absolute productivity measurement, but productivity change shifts the focus toward improvement, rather than deficits. The different connotation of results may render productivity change results more appropriate and persuasive in the NFP context.

Secondly, productivity change analysis can provide a solution to the problem of complex and messy data. Productivity change measurement not only shifts the focus of analysis to trends rather than levels, but also this shift precludes the necessity of having to estimate absolute levels of productivity at all. When data are complex and when organizations have multiple inputs and outputs, it may be difficult to determine a single absolute productivity ratio to encapsulate performance. Because productivity change measurement is concerned only with performance trajectories—that is, the shape of a curve between two points—the measurement doesn't actually require exact knowledge about where those two points sit relative to absolute zero. The implication is that assumptions imposed upon the data to aggregate multiple data elements do not carry quite as much weight. When the objective is only to understand the extent to which an organization has improved, rather than the extent to which one organization is more productive than another, methods to aggregate complex data can be made simpler. The relative simplicity of productivity change calculations will be demonstrated and made clearer in the chapters that follow. At this point, however, it is important to be aware that this is one of the reasons why an analyst might choose to employ a productivity change approach.

4.3   A Detailed Look at Measuring Productivity Change

4.3.1   Estimation Technique

Because there exists no standard unit of productivity, productivity measures nearly always require some type of transformation for interpretation. Recall in the previous chapter that we interpreted results by normalizing measured values with respect to the maximum observed value. Productivity change measurement involves comparing subsequent measures to an initial starting point. The method involves creating an index, where initial conditions are given a value of "1," and we track subsequent dynamics with reference to the base conditions.

4.3.2 Defining TI

The Törnqvist Index (TI)[1] is a standard productivity change measurement technique used by leading statistics and economics agencies around the world. It has most commonly been applied to measure changes in entire industries, sectors, or economies at large. More recently, however, researchers have espoused its use for productivity measurement in higher education. The TI's structure allows for transparent decomposition of results to understand the drivers of change, and its formula allows for the incorporation of multiple inputs and outputs.

A single TI measurement indicates a change in productivity from one time point to the next. The index produced is best understood in terms of percent change. For example, a single institution would need to have measures of inputs and outputs from at least two different points in time. If leaders want an estimate of how much productivity has increased or decreased over that period, they would feed all the input and output measures from both points in time into the TI formula, and they might get a number like "1.10" as a result. What this means is that productivity has increased by a factor of 1.1, or has increased by 10% over the period. Alternatively, if the TI score turned out to be 0.90, this would mean that productivity has decreased by a factor of 0.9, or has decreased by 10% over the period.

You can further think of the process as another example of feature scaling the data. Instead of normalizing our data with respect to a maximum value, as we did in Chapter 3, this time we are rescaling the data with respect to a magnitude of change over time, oriented around a base value of "1," representing no change.

The beauty of the TI is the systematic aggregation process it allows for, where the magnitude of change of different inputs and outputs can be weighted according to their relative value shares of total inputs and outputs. For example, with inputs that are valued monetarily, if Input A is worth ten dollars from a set of inputs worth 100 dollars in total, then the change index for Input A would be weighted 1/10 in terms of how much it contributes to the overall change of all inputs. Output weighting is treated the same. Thus, if input and output values are known empirically and precisely, as with monetary figures, then the TI works seamlessly. But even if input or output values are not known directly, as with teaching and research indicators, meaningful estimates can still be produced. The TI formula does not necessarily need exact figures to estimate value. It only requires estimates of value in terms of proportions of the whole, where the whole is always understood as 100%.

This may seem confusing, but we'll see how this works below in the tutorial. What is important to understand is the TI estimates can be easily decomposed for scrutiny and analysis, and even when only contextual information or human judgment is available to account for the value of inputs or outputs, TI measures can still generate robust estimates when modeling assumptions are made explicit.

---

[1] Caves, D. W., Christensen, L. R., & Diewert, W. . E. (1982). The Economic Theory of Index Numbers and the Measurement of Input , Output , and Productivity. Econometrica, 50(6), 1393–1414.

4.3.3   NRC Model (Adjusted Load)

As touched on in the previous chapter, the challenge of designating a productivity indicator for an education institution or NFP rests largely with how to characterize outputs. The output measure should encapsulate both the scale of service delivery and the value the service is producing without the luxury of using financial figures to indicate value. The US National Research Council (NRC) published a comprehensive report[2] outlining a rationale and technique for tracking both the quantity and value of educational outputs. They illustrate how past empirical research findings and TI calculation provide a solution. The NRC technique for measuring outputs requires counts of student course load and graduate completions. The entirety of an institution's student course load indicates the scale of the educational services being delivered at the institution. Completions indicate value that students gain from having earned a qualification. Further, because TI tracks productivity change over time, the ratio of completions to student load over time can give some indication of the effectiveness of educational programs and rates of student success.

Based on prior work[3] [4], the NRC argues that, on average, the value of being awarded a bachelor's degree is equivalent to the value of studying full-time in an institution for 1 year. The basis for the claim lies in an analysis of students' earnings after college relative to the amount of education they received. Research suggests that the bump in earnings from completing a qualification after 4 years of study is similar to the bump in earnings one might receive from completing 2 years of college instead of 1, or 3 years of college instead of 2. In other words, if we assume there is an inherent value in taking university courses (an assumption that should not be too farfetched), then the additional value of earning a qualification is similar to having studied for 1 year. The implication for measurement is that, if we understand equivalent full-time student load as an output measure for the scale of educational service delivery, then an undergraduate completion may be counted as one extra full-time-equivalent student—or in the case of US education an additional 32 credit-hours' worth of course delivery.

The NRC acknowledges how this metric does not hold for individual students in individual circumstances and further acknowledges how the estimation of value based on student earnings does not account for other types of value that students gain from a university education. They further note that the value of a degree likely differs from one institution to the next.

All these caveats, however, highlight the relevance and importance of using TI methods to track productivity change, rather than measures of absolute productivity. If we can agree that there exists some inherent value in taking university courses,

---

[2] Sullivan, et al. (2012). Improving Measurement of Productivity in Higher Education. National Academies Press, Washington, DC.

[3] Jaeger, D. A., & Page, M. E. (1996). Degrees matter: New evidence on sheepskin effects in the returns to education. The review of economics and statistics, 733-740.

[4] Park, J. H. 1999. Estimation of the sheepskin effect using the old and the new measures of educational attainment in the Current Population Survey. Economics Letters, 62: 237–240.

and there exists some inherent value in earning a qualification, then the TI will reveal the extent to which productivity for a single institution changes over time, relative only to that institution at the beginning of the period. Thus, we avoid the rathole of arguing why or why not one university should be portrayed as more or less productive in an absolute sense. If two universities' TI indexes are compared for 10 years of data, then the TI says nothing about the extent to which one institution is more productive than the other—only whether one university is on a better or worse trajectory for improving its productivity.

The TI is a relatively safe technique for baselining productivity. If institutional stakeholders have reason to believe that the absolute productivity of their institution is greater than that of another, then that argument could (and should) be made separately. The value of the TI and the NRC approach is to level the playing field for all institutions and use consistent measures that all universities track. It further uses well-accepted research to support assumptions about both quantity and value of outputs.

The purpose of this chapter is not to address all arguments for or against using particular measures and techniques. This section merely introduces one of the most defensible techniques available, based on the ability to deconstruct the metric and fully understand the mechanisms driving the end productivity portrayals the index produces. In later chapters, we will acquaint ourselves with another common technique that operates more as black box, relying on sophisticated, less-penetrable mathematics and introducing a new set of caveats.

4.3.4   Adjusting for the Real Value of Money

Creating a robust metric that estimates productive change over time should adjust for currency inflation when tracking financial indicators. Adjusting for inflation is one way to better represent the "real value of money" over a period. The consumer price index (CPI) and GDP price deflators are commonly used in economics. More precise monetary adjustments can be made for individual sectors or industries. The NRC model uses financial measures to represent inputs. They include employee expenditures, capital expenditures, and other intermediate expenditures. The term "intermediate" signifies operational costs that cannot be considered direct educational expenditures.

To not adjust for the real value of money when examining the productivity of an individual institution means capturing broad fluctuations in the economy, rather than capturing changes in only the "machinery" and productive processes of that single institution. To illustrate the importance of making the adjustment over a period, imagine that monetary inflation is about 2% per annum. Over 10 years, the cumulative inflation would be about 22%. If we are interested in calculating institutional productivity over the same period, then we want to factor that 22% out of the equation. If we do not, then inputs will appear to be increasing more quickly than they should, relative to outputs. In the case of the NRC model, with monetary inputs and non-monetary outputs, if you do not adjust for currency inflation, then you will inherently underestimate productivity because monetary input growth will be naturally positioned to outpace non-monetary output growth. Further, it is standard practice in any economic productivity analysis to adjust for currency inflation when examining time series data. Either way, one should have a clear rationale for making the adjustment or for not making the adjustment. The implications should be understood and not overlooked.

4.4 Tutorial: Measuring University Productivity Change

4.4.1 Setup

This tutorial requires the following packages. We introduce two new packages that help to keep our charts tidy as we will be creating more of them in this chapter.

```
library(knitr)
library(tidyverse)
library(ggrepel) # prevents overlapping text in charts
library(scales)  # easy manipulation of chart labels
```

4.4.2 Introduction

Higher education system leaders are interested to understand high-level implications of different educational service delivery models at different universities. They wish to observe how the dynamics play out over time. They selected four top-performing institutions in different states to examine. They decided they want to use the US NRC model as their basis for analysis. Higher education reporting requirements in each state require these institutions to report expenditures toward different educational functions separately. Salaries, capital, and intermediate and operational expenditures must be given for instruction and research separately. This accounting system, while imperfect, allows for better linking of inputs and outputs for different types of academic work.

4.4.3 Get the Raw Data

The data below represent the inputs and outputs of the four selected institutions over 12 years. Data are given for each institution at equal time intervals across the period. The outputs are full-time equivalent students based on the institution's overall course load (credit hours) and the number of graduate completions (qualifications awarded) for the indicated year. The inputs are separated into three categories, salaries, capital expenditure, and intermediate. The inputs are the institution's reported expenditures on instruction only.

```
# cut this down. to only coursework FTE and completions for the
outputs. break out the expenditures to standard categories. Change
to anonymous uni names and round figures.


uni_data_edu <- tribble(
    ~institution, ~year, ~x_salaries, ~x_cap_ex, ~x_intermediate,
~y_fte_course, ~y_compl_course,
  "University A",   2011,    12160000,   3760000,        5840000,
20900,           3940,
  "University A",   2013,    13000000,   4800000,        7200000,
23640,           4554,
  "University A",   2015,    16560000,   5320000,        8880000,
25580,           6414,
  "University A",   2017,    19400000,   7320000,       10280000,
26980,           6224,
  "University A",   2019,    20040000,   7520000,       10920000,
29600,           6444,
  "University B",   2011,     6350000,   1050000,        4000000,
9830,            2122,
  "University B",   2013,     7350000,   1450000,        4300000,
12080,           1737,
  "University B",   2015,     9450000,   1050000,        4450000,
13510,           2680,
  "University B",   2017,    10200000,   1350000,        4900000,
16350,           3585,
  "University B",   2019,    10700000,   1200000,        4500000,
15100,           3402,
  "University C",   2011,    10840000,   2120000,        5600000,
21540,           5580,
  "University C",   2013,    12440000,   2600000,        6800000,
23775,           5590,
  "University C",   2015,    15440000,   3600000,        8200000,
26745,           6015,
  "University C",   2017,    17120000,   4280000,        8680000,
28035,           6337,
  "University C",   2019,    18840000,   4240000,       10520000,
28935,           6648,
  "University D",   2011,     5500000,    550000,        2300000,
12290,           2855,
  "University D",   2013,     6350000,    650000,        3150000,
14950,           3060,
  "University D",   2015,     7000000,    700000,        3100000,
16070,           3350,
```

```
   "University D",   2017,    8050000,    750000,         3200000,
18360,              3782,
   "University D",   2019,    9000000,   1100000,         3700000,
20170,              4420
   )
```

4.4.4   Adjust for RVM and Adjusted Load

We now want to make appropriate adjustments to the data for our productivity calculation. The first is to use CPI information to adjust for the real value of money
over the period. We will convert all monetary figures into 2007 dollars. Based on
prior research, we can use the deflator values below. The deflator vector below gives
fractional values to be multiplied against all monetary figures for the corresponding
years. The data show that by 2019, currency inflation has caused the value of money
to drop, such that one dollar can buy only about 81% of what it could purchase
in 2011.

```
# define rvm based on research
rvm_2007 <- tibble(
  year = c(2011, 2013, 2015, 2017, 2019),
  rvm = c(1.000, 0.952, 0.896, 0.850, 0.812)
)

# adjust all monetary values for each institution to 2007 dollars
unis_rvm_edu <- uni_data_edu %>%
  group_by(institution) %>%
  mutate(across(starts_with("x"),  # performs function below on
each column beginning with "x"
               function(var) var*rvm_2007$rvm)) %>%
  ungroup()
```

In the second chunk of code above, we group by institution, so R knows to multiply the RVM vector for each institution four separate times—that is, for every time
each year occurs. The mutate_at() allows us to denote a condition, so that the operation is performed on each column satisfying the condition. In this case, we designate
each input column containing financial figures. If we examine the table below, we

Table 4.1 Financial figures adjusted for the 'real value of money', base year 2011

Institution	Year	x_salaries	x_cap_ex	x_intermediate
University A	2011	12160000	3760000	5840000
University A	2013	12376000	4569600	6854400
University A	2015	14837760	4766720	7956480
University A	2017	16490000	6222000	8738000
University A	2019	16272480	6106240	8867040

can see the effect. The new adjusted financial figures for University A are shown, so you may compare with the raw data given above (Table 4.1).

```
unis_rvm_edu[1:5, 1:5]
```

If we examine the data in the new table, we can see that for each institution, the 2011 financial values have remained unchanged, and the values of subsequent have been adjusted to reflect the effects of inflation. We do this because we want to capture only true changes in the productivity of the institution. That is, we want to measure genuine differences in the institutional production function, and we want to filter out fluctuations in the broader economy.

The next step is to follow NRC methods to construct our output variable. We do this by mutating a new column, named y_adjusted_load, which is the sum of y_fte_course and y_compl_course. The new variable captures both the scale of educational operations and the value of completing a qualification.

```
# calculate adjusted load
unis_adjusted_edu <- unis_rvm_edu %>%
  mutate(y_adjusted_load = y_fte_course + y_compl_course)

glimpse(unis_adjusted_edu)
## Rows: 20
## Columns: 8
## $ institution       <chr> "University A", "University A",
"University A", "Un...
## $ year              <dbl> 2011, 2013, 2015, 2017, 2019, 2011, 2013,
2015, 201...
## $ x_salaries        <dbl> 12160000, 12376000, 14837760, 16490000,
16272480, 6...
## $ x_cap_ex          <dbl> 3760000, 4569600, 4766720, 6222000,
6106240, 105000...
## $ x_intermediate    <dbl> 5840000, 6854400, 7956480, 8738000,
8867040, 400000...
## $ y_fte_course      <dbl> 20900, 23640, 25580, 26980, 29600, 9830,
12080, 135...
## $ y_compl_course    <dbl> 3940, 4554, 6414, 6224, 6444, 2122,
1737, 2680, 358...
## $ y_adjusted_load   <dbl> 24840, 28194, 31994, 33204, 36044,
11952, 13817, 16...
```

Now save this data frame for use later. Recall that you must first create a folder in your working directory, called "data" to prevent getting an error.

```
save(unis_adjusted_edu, file = "data/unis_adjusted_edu.rda")
```

4.4.5   Calculate Change Indexes

Creating change indexes serves two purposes. The first is to transform our data from static values at single points in time to more dynamic values that capture the magnitude of change from one time point to the next. The second purpose is to convert all data elements into a common unit of measure. Once all data elements are expressed as factors of change, they can be aggregated in a more straightforward and robust way. If we want to see overall trends incorporating elements measured in different units, the change indexes largely solve the problem of adding together apples and oranges.

The objective of this section is to transform our data from static values. Calculate change indexes for each element. The next step is to put our data into "long" format. This will allow us to make all the calculations we need to make an all variable, simultaneously. Not only does this make our scripts more concise and readable, but also it's faster for the computer! Above we needed to make a transformation to only the input variables, so we used the mutate_at function to designate the appropriate variables. From here, though, we want to perform the same change index calculation for all universities and all inputs and outputs. Long format will help us achieve that.

```
unis_adjusted_long <- unis_adjusted_edu %>%
  pivot_longer(cols = x_salaries:y_adjusted_load,
               names_to = "variable",
               values_to = "value") %>%
  mutate(nature = ifelse(grepl("x_", variable), "x", "y")) %>%
  select(institution, variable, nature, year, value) %>%
  arrange(institution, variable, year)

glimpse(unis_adjusted_long)
## Rows: 120
## Columns: 5
## $ institution <chr> "University A", "University A", "University
A", "Univer...
## $ variable    <chr> "x_cap_ex", "x_cap_ex", "x_cap_ex", "x_cap_
ex", "x_cap_...
## $ nature      <chr> "x", "x", "x", "x", "x", "x", "x", "x", "x",
"x", "x", ...
## $ year        <dbl> 2011, 2013, 2015, 2017, 2019, 2011, 2013,
2015, 2017, 2...
## $ value       <dbl> 3760000, 4569600, 4766720, 6222000, 6106240,
5840000, 6...
```

We will employ lag() function from the dplyr package within the tidyverse library. When using dplyr functions with %>%, we are frequently operating on entire columns of data at a time. And when we have data in long format, we are operating on only one column that contains many variables. R, however, is still producing element-by-element results. Within a piped chunk of code, lag() tells R to call the immediately preceding element in the column to where it would normally be operating.

Let's see a quick example to illustrate the functionality. Divide a vector v by lag(v), and see what happens.

```
v <- c(1,2,4,8)
v/lag(v)
## [1] NA   2   2   2
```

Our output is a vector of the same length as v, where the first element is NA. This is because there is no previous element to divide by in the first instance. The next three values are the result of taking v[2]/v[1], v[3]/v[2], and v[4]/v[3], or 2/1, 4/2, and 8/4, respectively.

We can see how handy this will be now that we have our data in long format, ordered by subsequent years. Because we're going to be using the above operation repeatedly, let's create a concise function for calculating the change factor of a variable from one time point to the next using lag(). We create this function because we'll be using it throughout this chapter and the next, and it helps make our code more readable and less redundant.

We need only make one addition to the example above. We need to replace the NA value with a value of 1. Recall our objective is to create an index relative to a designated starting point, and 1 gives us a standard frame of reference.

```
# create function
change_factor <- function(col) {

  new_col <- col/lag(col)
  new_col[1] <- 1

  return(new_col)
  }
```

This function will come in handy in later chapters, so let's save it for easy access.

```
save(change_factor, file = "data/change_factor.rda")
```

We can now put our new function to work. Let's operate on unis_adjusted_long to generate a new column with change indexes for each data element. First, we must group by institution and variable, so R knows to perform the function separately for each data element for each university across the period. We can then take a sample of the new data to inspect with the change indexes look like.

```
uni_change_indexes <- unis_adjusted_long %>%
```

```
group_by(institution, variable) %>%
mutate(change_index = change_factor(value)) %>%
ungroup()
```

```
glimpse(uni_change_indexes)
## Rows: 120
## Columns: 6
## $ institution  <chr> "University A", "University A", "University
A", "Unive...
## $ variable     <chr> "x_cap_ex", "x_cap_ex", "x_cap_ex", "x_cap_
ex", "x_cap...
## $ nature       <chr> "x", "x", "x", "x", "x", "x", "x", "x",
"x", "x", "x",...
## $ year         <dbl> 2011, 2013, 2015, 2017, 2019, 2011, 2013,
2015, 2017, ...
## $ value        <dbl> 3760000, 4569600, 4766720, 6222000, 6106240,
5840000, ...
## $ change_index <dbl> 1.0000000, 1.2153191, 1.0431373, 1.3053001,
0.9813950,...
```

Now select one of the data elements from one of the universities to get a sense for what these change indexes are telling us. We'll plot the data using a line chart and add a reference for our base value of 1 (Fig. 4.1).

```
uni_change_indexes %>%
  filter(institution == "University A",
         variable == "x_cap_ex") %>%
  ggplot(aes(x = year, y = change_index)) +
```

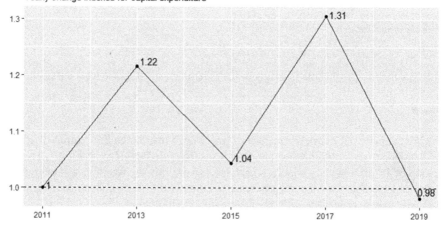

Fig. 4.1 The noisy results of year-on-year productivity change scores

```
geom_line() +
geom_point() +
geom_hline(yintercept = 1, linetype = "dashed") +
geom_text_repel(aes(label = round(change_index, 2))) +
scale_x_continuous(breaks = seq(2011, 2019, by = 2)) +
labs(y = "",
     x = "",
     subtitle = "Yearly change indexes for capital expenditure")
```

Interpretation of the chart above may not be immediately intuitive. Any points above the y = 1 reference line signify a positive change from the previous observation. Anything below the line represents a negative change. The dip in 2015 does not signify a decrease. It signifies only a reduction in the rate of increase. In 2013 capital expenditure increased by 22% from 2011. In 2015 it increased by 4% from its 2013 value. The only decrease in capital expenditure occurred in 2019, where it was only 98% of what it was in 2017, i.e., a 2% decrease.

4.4.6   Calculate Cumulative Change

If you Google multi-factor productivity trends from your government's bureau of statistics, you will likely find trend lines resembling the one above. These charts are helpful when you are most interested in the magnitude of the incremental change between years. They are less helpful for casual, less discriminating interpretation. This is why we want to create a second index that captures the cumulative change from the base year across the period. We want to easily answer the question, by what factor has each data element changed in total by 2019?

To calculate this, we use R's cumulative product function cumprod(). To illustrate how it works, let's take the cumulative product of our vector v from above.

```
v <- c(1,2,4,8)
cumprod(v)
## [1]  1  2  8 64
```

cumprod() also returns a vector the same length as the original vector. The following chunk of code illustrates exactly what's happening, so we can see how using this function on our change indexes will give us the cumulative change across the period

```
cumprod(v)  ==  c(v[1],  v[1]*v[2],  v[1]*v[2]*v[3],  v[1]*v[2]*
v[3]*v[4])
## [1] TRUE TRUE TRUE TRUE
```

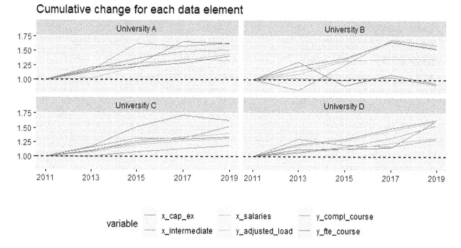

Cumulative change for each data element

Fig. 4.2 The more intuitive and predictable trends of 'cumulative change' scores

Let's use this function now to calculate the cumulative change of each element for each university. We can then plot all results to get a change profile for all four universities (Fig. 4.2).

```
# calculate cumulative change
uni_elements_cum_change <- uni_change_indexes %>%
  group_by(institution, variable) %>%
  mutate(cum_change = cumprod(change_index)) %>%
  ungroup()

# plot
uni_elements_cum_change %>%
  ggplot(aes(x = year, y = cum_change, color = variable)) +
  facet_wrap(~institution) +
  geom_line() +
  geom_hline(yintercept = 1, linetype = "dashed") +
  scale_x_continuous(breaks = seq(2011, 2019, 2)) +
  theme(legend.position = "bottom") +
  labs(y = "",
       x = "",
       title = "Cumulative change for each data element")
```

The chart above shows that for Universities A, C, and D, capital expenditures have been on the rise across the entire period, but University D has decreased its capital expenditure overall. Most institutions have also been increasing their rate of course completions, but University C's course completions are not increasing at the rate of other key variables.

4.4.7   Further Functionalize Our Code

Calculating change indexes and cumulative change is something we'll be carrying out repeatedly. So we'll wrap both these calculations up in a single function. This is good practice in complex analyses, both cut down on the amount of code we'll be writing, and also to force consistency and reduce the potential for errors. You'll see further below in a subsequent chunk of code how it simplifies matters.

The objective for this new function is to operate like any major function we can just incorporate into our piping from the original data frame. It will create new columns with consistent naming, and we won't have to explicitly call group_by() and mutate() so many times. It should thus take four prime arguments: (1) the data frame on which we're operating, (2) the column of values from which we want to calculate our index, (3) a specification for calculating either change indexes using change_factor() or cumulative change using cumprod(), and (4) the appropriate grouping variables.

The code below uses some notation we have not seen before. From an R programming perspective, it contains the most technical code that we'll see in this book. The reason for the technicality is to allow for flexibility in use. No matter the task, if you need to create a change index relative to a base year, you'll be able to copy and paste this code and use it flexibly. Each chunk is commented for easier interpretation. We'll get to see the new function in action and how much it simplifies our script in the next section.

```
# specify function arguments
calc_change <- function(df, var, calc, ...) {

  # store the grouping variables so they can be used
  # within the group_by() function below.
  group_vars <- quos(...)

  # specify the calculation to use and the name of the new column
  # if "change_index" is specified in the function argument
  if(calc == "change_index"){
    FUN <- change_factor
    new_var_name <- paste(var, "index", sep = "_")

  # specify the calculation to use and the name of the new column
  # if "cum_change" is specified in the function argument
  } else if(calc == "cum_change"){
    FUN <- cumprod
    new_var_name <- paste("cum_change", var, sep = "_") %>%
    str_remove("_index")
  }

  # use the grouping variables and the new variable names
```

```
# for calculating the specified change index
# special notation is used to use the stored grouping variables
# and assign the new variable name within the function
new_data <- df %>%
    group_by(!!!group_vars) %>% # notation for calling grouping
variables
    mutate(!!new_var_name := FUN(.data[[var]])) %>%
    ungroup()

    return(new_data)
}


# save
save(calc_change, file = "data/calc_change.rda")
```

4.4.8   Calculate Data for TI Weights

We've now demonstrated how to calculate change indexes and cumulative change of individual data elements. We now proceed to aggregate the data to produce TI productivity change indicators for the four institutions in our dataset and to utilize the calc_change() function created above. As discussed above, individual input (or output) data elements in a TI calculation are weighted according to their respective value shares of the total value of inputs (or outputs). We thus need to calculate those value shares as proportions from our original data. The table below serves to clarify what is being accomplished in this step.

We'll begin the TI calculation process with that long data we created above, unis_adjusted_long. The chunk of code below first filters out redundant data and then calculates the value shares of each element for each institution in each year. Regarding the filter, we first need to exclude y_compl_course and y_fte_course before calculating proportions and creating our TIs. One of the purposes of employing the NRC model to create y_adjusted_load was to reduce the output elements in the TI calculation to one. We'll explore how to manage multiple output elements in the next chapter.

```
# create a couple small functions
# prop_calc <- function(var) var/sum(var)
```

Table 4.2 Lagged means of value shares

Institution	Variable	Nature	Year	Value	Share	lagged_share_mean
University A	x_cap_ex	x	2011	3760000	17.3%	17.3%
University A	x_cap_ex	x	2013	4569600	19.2%	18.2%
University A	x_cap_ex	x	2015	4766720	17.3%	18.2%
University A	x_cap_ex	x	2017	6222000	19.8%	18.5%
University A	x_cap_ex	x	2019	6106240	19.5%	19.7%

```
# lagged_mean <- function(var)  (var + lag(var))/2
# filter and calculate element value shares
# per university per year
prod_data_long <- unis_adjusted_long %>%
   filter(!variable %in% c("y_compl_course", "y_fte_course")) %>%
   group_by(institution, year, nature) %>%
   mutate(share = value/sum(value)) %>%
   ungroup()


# check data for one institution in one year
prod_data_long %>%
        filter(institution == "University A",
               year == 2011) %>%
        mutate(share = percent(share, accuracy = .1))
## # A tibble: 4 x 6
##    institution  variable        nature  year     value share
##    <chr>        <chr>           <chr>   <dbl>     <dbl> <chr>
## 1 University A x_cap_ex         x        2011  3760000 17.3%
## 2 University A x_intermediate   x        2011  5840000 26.8%
## 3 University A x_salaries       x        2011 12160000 55.9%
## 4 University A y_adjusted_load y         2011     24840 100.0%
glimpse(prod_data_long)
## Rows: 80
## Columns: 6
## $ institution <chr> "University A", "University A", "University
A", "Univer...
## $ variable    <chr> "x_cap_ex", "x_cap_ex", "x_cap_ex", "x_cap_
ex", "x_cap_...
## $ nature      <chr> "x", "x", "x", "x", "x", "x", "x", "x", "x",
"x", "x", ...
## $ year        <dbl> 2011, 2013, 2015, 2017, 2019, 2011, 2013,
2015, 2017, 2...
## $ value       <dbl> 3760000, 4569600, 4766720, 6222000, 6106240,
5840000, 6...
## $  share      <dbl> 0.17279412, 0.19200000, 0.17295189,
0.19783784, 0.19542...
```

The table above makes clear that salaries, for example, represent about 56 percent of the value of all inputs. Adjusted load, as our only output, represents 100
percent of the value of outputs.

   We now find the average value share between each pair of subsequent timepoints
in the dataset. Recall from our calculations above, each change index incorporates
information from both time t and time $t-1$ using the lag() function. Since we'll be
using the value shares as weights for the change indexes, we also need to use the
lag() function to find the average share for each element across each intermittent

period of change. We also need to replace the NA values produced by the lag() function similar to above (Table 4.2).

```
# calculate average value shares and replace NA
# values in 2011 with the
prod_data_long <- prod_data_long %>%
  group_by(institution, variable) %>%
  mutate(lagged_share_mean = (share + lag(share))/2,
          lagged_share_mean = ifelse(is.na(lagged_share_mean),
share, lagged_share_mean)) %>%
  ungroup()


# reexamine data for one institution in one year
prod_data_long %>%
  filter(institution == "University A",
         variable == "x_cap_ex") %>%
  mutate(across(contains("share"), percent, accuracy = .1))
```

Now that we have all the information we need for aggregation of the change indexes, we can now demonstrate the use of the function we created above to simply produce those change indexes.

```
# create change indexes
prod_data_long <- prod_data_long %>%
  calc_change(var = "value",
              calc = "change_index",
              institution, variable)

# examine all new data
glimpse(prod_data_long)
## Rows: 80
## Columns: 8
## $ institution        <chr> "University A", "University A",
"University A", "...
## $ variable           <chr> "x_cap_ex", "x_cap_ex", "x_cap_ex",
"x_cap_ex", "...
## $ nature             <chr> "x", "x", "x", "x", "x", "x", "x",
"x", "x", "x",...
## $ year               <dbl> 2011, 2013, 2015, 2017, 2019, 2011,
2013, 2015, 2...
## $ value              <dbl> 3760000, 4569600, 4766720, 6222000,
6106240, 5840...
## $ share              <dbl> 0.17279412, 0.19200000, 0.17295189,
0.19783784, 0...
```

```
## $ lagged_share_mean <dbl> 0.17279412, 0.18239706, 0.18247594,
0.18539486, 0...
## $ value_index         <dbl> 1.0000000, 1.2153191, 1.0431373,
1.3053001, 0.981...
```

4.4.9   Calculate TIs

Now that we have all the information we need to aggregate the individual element change indexes, we can calculate composite input and output TIs. The process involves taking the product of the change indexes raised to the power of their weights—and we do this separately for inputs and outputs. The order of operations is to first raise each change index to the power of its weight, the mean value share. Then for each institution in each year, we aggregate the component inputs and outputs, i.e., we take the product of all weighted inputs and the product of all weighted outputs. In this example, there is only one output, so there is no aggregation necessary.

```
# calculate composite indexes
composite_indexes <- prod_data_long %>%
  mutate(weighted_index = value_index^lagged_share_mean) %>%
  group_by(institution, nature, year) %>%
  summarise(composite_index = prod(weighted_index)) %>%
  ungroup()
```

```
# inspect new data
glimpse(composite_indexes)
## Rows: 40
## Columns: 4
## $ institution     <chr> "University A", "University A",
"University A", "Un...
## $ nature          <chr> "x", "x", "x", "x", "x", "y", "y", "y",
"y", "y", "...
## $ year            <dbl> 2011, 2013, 2015, 2017, 2019, 2011, 2013,
2015, 201...
## $ composite_index <dbl> 1.0000000, 1.0937600, 1.1580056,
1.1411472, 0.99350...
```

The table above shows how we aggregated all component change indexes into only two composite input and output indexes for each year. The final step is to calculate the productivity change index by dividing the output indexes by the input indexes for each year. The method we will use is to temporarily pivot our data into wide format, divide the output variable by the input to create a new productivity variable, and then pivot back to long format so we can then use our calc_change function to produce cumulative change indexes to make interpretation of results more intuitive.

```
# calculate productivity change
prod_change_edu <- composite_indexes %>%
    pivot_wider(names_from = "nature", values_from = "composite_
index") %>%
  mutate(P = y/x) %>%
    pivot_longer(x:P, names_to = "nature", values_to = "composite_
index") %>%
  arrange(institution, nature, year) %>%
  calc_change(var = "composite_index",
              calc = "cum_change",
              institution, nature)

# view results
glimpse(prod_change_edu)
## Rows: 60
## Columns: 5
## $ institution          <chr> "University A", "University A",
"University A"...
## $ year                 <dbl> 2011, 2013, 2015, 2017, 2019, 2011,
2013, 2015...
## $ nature               <chr> "P", "P", "P", "P", "P", "x", "x",
"x", "x", "...
## $ composite_index      <dbl> 1.0000000, 1.0377269, 0.9799438,
0.9094529, 1....
## $ cum_change_composite <dbl> 1.0000000, 1.0377269, 1.0169141,
0.9248355, 1....
```

Save results for use later.

```
save(prod_change_edu, file = "data/prod_change_edu.rda")
```

Let's now plot the data to see the productivity trends and to understand the dynamics produced from input and output growth. In the next chapter, we'll be plotting productivity change in the next chapter as well, so let's wrap up this plot into a function and save it.

```
# visualise
plot_prod_change <- function(df) {

  # set up chart parameters
  plot <- df %>%
```

```
  ggplot(aes(x = year, y = cum_change_composite, color = nature,
size = nature)) +
    facet_wrap(~institution) +
    geom_line() +
    scale_size_manual(values = c(1.5, 0.5, 0.5)) +
    geom_hline(yintercept = 1, linetype = "dashed") +

    # adjust the cosmetics
    scale_x_continuous(breaks = seq(2011, 2019, by = 2)) +
    labs(x = "",
         y = "",
         color = "",
           subtitle = "Cumulative productivity, input and output
change by institution") +
      scale_color_discrete(labels = c("Productivity", "Inputs",
"Outputs")) +
    theme(legend.position = "bottom") +
    guides(size = FALSE)

  return(plot)
}


save(plot_prod_change, file = "data/plot_prod_change.rda")
```

Now use the function on our productivity change data set (Fig. 4.3).

```
plot_prod_change(prod_change_edu)
```

Positive productivity change trends depend upon output growth exceeding input growth. Universities B and D each have consistently positive productivity trends, and we can see that the rate of growth of the outputs remains consistently greater than that of the inputs.

4.5   Reflections on Productivity Change in Practice

Productivity change benchmarks generate insight about performance trajectories. They reveal the dynamics of progress or decline or highlight the magnitude of a shock or turning point for an institution. They indicate nothing about whether one institution's productivity is greater than another's. Analysts using TIs must exercise restraint in forming conclusions across institutions. A university's initial conditions are baked into the portrayal of its productivity change. While TIs produce robust and transparent measures of performance over time, their change portrayals can be misleading. Using TIs in practice requires acknowledging variation across a field of institutions in their base levels of quality and performance.

Cumulative productivity, input and output change by institution

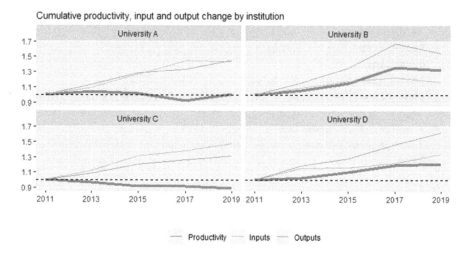

Fig. 4.3 Cumulative poductivity trends vs. input and output trends

TIs are most powerful for operations research and analysis when comparisons of performance levels are less important. Consider how TIs have been used historically, that is, indicating overall economic trends. What would it mean necessarily to find that the oil and gas industry is three times more efficient at transforming inputs into outputs than the agriculture industry? Such a finding might only cement the understanding that the natures of these two industries are different. Alternatively, a far more critical finding would be to learn that a country's agricultural productivity is increasing threefold every 10 years, while productivity in oil and gas is remaining flat. We now immediately begin to ask questions about the nature of innovation in agriculture versus innovation in oil and gas. Framing problems and issues is now more urgent, and questions about interventions or incentives begin to arise.

The same goes for productivity change measurement in higher education. Would it necessarily be important or surprising to find that Harvard University spends more per undergraduate completion than your local state institution? What if alternatively, you found that, for a common measure of change, the productivity of the private university is increasing twice as fast as that of the public one? Again, we now begin to ask questions about what is the nature of innovation (or austerity)—rather than inherent qualitative differences between private endowments and public budgets.

A fundamental requirement and assumption of productivity change analysis with TIs is that qualitative differences between institutions exist, and further, they might be extreme and difficult to measure. TIs help individual institutions better understand the extent to which their own internal "machinery" is changing relative to itself. Changes may also be related to external hindering or facilitating factors outside the control of the institution, but the TI will help communicate how well the institution is managing external pressures. If peer institutions are included in the analysis, they may indicate the relative effectiveness of different institutions' innovation and adaptation strategies, even if they cannot communicate overarching levels of productivity and performance.

Productivity Change with Threshold Analysis

5

5.1 Objective

This chapter explores TI methods for incorporating, not only multiple types of inputs but also multiple types of outputs. By the end of this chapter, the reader will be able to handle increasingly complex datasets and generate alternative productivity portrayals of institutions over time. The chapter extrapolates on the concept of adjusted student load to define an analogous research output measure, "adjusted publications." This chapter helps the reader acquire a sharpened sense for aggregating multiple indicators without losing information or obscuring the dynamics of individual data elements. The reader will walk away with a new and bespoke analytical technique, threshold analysis, to explore a range of plausible interpretations of observed trends.

5.2 What Is Threshold Analysis?

A threshold analysis of productivity change establishes a plausible range of performance portrayals that could arise from single set of data.[1] The bounds on results reflect stakeholder values and expert value judgments. The analysis produces alternative performance trajectories over time, based on different ideas about what is important to emphasize in the data. Stakeholders often have differing interests in how data is collected, used, and reported. Threshold analysis attempts to reconcile these interests. The purpose of threshold analysis is to set bounds on the subjectivity involved in both the use and interpretation of data. The figure below demonstrates what the output of a threshold analysis may look like (Fig. 5.1).

---

[1] Moore, K., Coates, H., & Croucher, G. (2019). Investigating applications of university productivity measurement models using Australian data. Studies in Higher Education, 44(12), 2148-2162.

© The Author(s), under exclusive license to Springer Nature
Switzerland AG 2021
K. Moore, *Measuring Productivity in Education and Not-for-Profits*,
Management for Professionals, https://doi.org/10.1007/978-3-030-72965-3_5

Range of plausible productivity trends

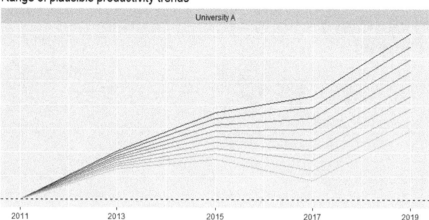

Fig. 5.1 An example visualisation of threshold analysis

Threshold analysis provides a minimally controversial approach for assigning and testing different value weights for various data elements. It achieves this in two ways. First, the analysis is not intended to reach definitive conclusions, but rather present a bounded set of options with assumptions and implications. Second, output value weights are assigned proportionally, not directly. That is, no heroic assumptions are required for precise valuation of any university outputs—such as a degree or a research publication—in monetary terms or otherwise. The beauty of using TIs for threshold analysis is that when the precise value of outputs is not known or difficult to quantify, analysis can still proceed under assumptions that different outputs generate different proportions of the total value being produced.

5.3   A Detailed Look at Threshold Analysis for Universities

5.3.1   Multiple Outputs

Threshold productivity analysis becomes necessary when dealing with a dataset that contains multiple variables of fundamentally different type. The previous chapter incorporated two related output variables, student load and student completions, into a single output variable, called "adjusted load." Universities, however, perform multiple academic functions that are not centered around students. Research is the prime example. If we are interested of the productivity of an entire institution, we need to be able to account for the productivity of both education and research.

What would a single, aggregated education-research output variable look like? What would be its unit of measure? These are difficult questions. A workable solution, however, has been demonstrated in the literature using TIs. The remainder of this chapter will focus on that practical application.

5.3.2   Characterizing Production Technology

Distinguishing between different types of "production technology" can further illustrate the value of threshold analysis for understanding the performance among organizations with multiple outputs. We distinguish between "separate production" and "joint production." Separate production describes processes where different outputs derive from mutually exclusive sets of inputs. If the organization were to cease production of one of the outputs, then production of the other output(s) could continue unaffected. Joint production, on the other hand, describes processes where inputs are shared and the production processes are more difficult to examine in isolation. Take education and research as the example. University laboratories may be used for scientific experiments or for practical demonstrations for students. On the same day, one professor might write a journal article and also prepare a lesson plan. There also exists the idea of the "teaching-research nexus," where excellence in either domain is supposed to bolster the quality of the other. Finally, separating funding between the two functions proves difficult. Even though institutions track exact numbers of graduate completions and exact numbers of journal articles published, breaking down financial statements into teaching-only and research-only expenditures often proves difficult.

Threshold analysis enables a broader perspective on how well joint production processes are working. Because the production processes themselves are known to be hazy, analytical results leave room for interpretation. Using known parameters and contextual knowledge of an institution, threshold analysis allows an analyst to quantify the gaps and implications that may arise from different assumptions about outputs and the manner in which they are produced.

5.3.3   Adjusted Publications

In constructing the threshold analysis, we will define a new research output indicator using bibliometric data on university research publications. We call the new indicator "adjusted publications"[2]. The adjusted publications indicator attempts to capture the value of published research in a manner similar to how the "adjusted load" indicator captures the value of taking courses and completing a degree.

The adjusted publications indicator accounts for the additional value-add of a research publication's field-weighted citation impact. Essentially, articles that generate more citations are assumed to produce more inherent value. Examples from research[3] have assigned more value to scholarly articles ranked within the top 25% of highly cited articles within their respective fields. Empirical findings on the level of impact of highly

---

[2] Moore. K. (2019). Investigating higher education productivity and its measurement in Australia. Melbourne Graduate School of Education. Available at https://minerva-access.unimelb.edu.au/handle/11343/238674. Accessed 30 Nov. 2020.

[3] Yang, et al. (2020). The Productivity of Leading Global Universities: Empirical Insights and Implications for Higher Education. In Responsibility of Higher Education Systems (pp. 224-249). Brill Sense.

cited articles indicate that the publications in the top 25% generate about twice the impact of the average publication.[4] Hence, when accounting for scholarly publications as an output indicator, the articles in the top 25% are counted twice.

Both the adjusted load and adjusted publications output indicators are crude measures. But if we can agree that (A) there is more value in a student completing a degree than dropping out and that (B) there is more value in an article being read and cited more frequently than less, then we are simply establishing a consistent baseline for comparison. And recall that productivity change metrics are only measuring the steepness of productivity improvement trajectories relative to themselves. We are not using these measures to assert that one institution is definitively more productive than another.

5.4   Tutorial: Teaching-Research Nexus or Divide?

5.4.1   Setup

This tutorial requires several objects from the previous chapter. After activating the packages, we need the unis_adjusted_edu dataset three functions for calculating and plotting productivity change. You will need to have run the script from the previous chapter and saved all required objects for the following chuck of code to work.

```
# packages
library(knitr)
library(tidyverse)
library(glue) # intuitive functions for combining pieces of text

# data
load("data/unis_adjusted_edu.rda")

# functions
load("data/change_factor.rda")
load("data/calc_change.rda")
load("data/plot_prod_change.rda")
```

5.4.2   Introduction

Higher education system leaders suspect that incentives to emphasize research performance have overshadowed incentives to improve teaching and instruction. Drivers of the phenomenon may include international rankings systems that heavily weight institutional research output, as well as academic career progression practices within

---

[4] Moore. K. (2019). Investigating higher education productivity and its measurement in Australia. Melbourne Graduate School of Education. Available at https://minerva-access.unimelb.edu.au/handle/11343/238674. Accessed 30 Nov. 2020.

institutions that emphasize individual research performance. Leaders are questioning whether the balance has shifted too far away from teaching and learning. Our job is to provide a baseline of evidence to either support or challenge the claim.

5.4.3   Create New Data Set

We begin with research data from the four universities we examined in the previous chapter.

```
# use updated data


uni_data_res <- tribble(
     ~institution, ~year, ~x_salaries, ~x_cap_ex, ~x_intermediate,
~y_pubs_total, ~y_pubs_top25,
   "University A",  2011,    18240000,   5640000,        8760000,
3885,          1615,
   "University A",  2013,    19500000,   7200000,       10800000,
4728,          2073,
   "University A",  2015,    24840000,   7980000,       13320000,
6027,          2770,
   "University A",  2017,    29100000,  10980000,       15420000,
7297,          3615,
   "University A",  2019,    30060000,  11280000,       16380000,
8083,          4020,
   "University B",  2011,     6350000,   1050000,        4000000,
3025,          1253,
   "University B",  2013,     7350000,   1450000,        4300000,
3336,          1460,
   "University B",  2015,     9450000,   1050000,        4450000,
3803,          1744,
   "University B",  2017,    10200000,   1350000,        4900000,
4050,          1768,
   "University B",  2019,    10700000,   1200000,        4500000,
4359,          1960,
   "University C",  2011,    16260000,   3180000,        8400000,
2695,          1138,
   "University C",  2013,    18660000,   3900000,       10200000,
3301,          1408,
   "University C",  2015,    23160000,   5400000,       12300000,
4032,          1693,
   "University C",  2017,    25680000,   6420000,       13020000,
5271,          2415,
```

```
  "University C",   2019,     28260000,   6360000,        15780000,
5856,            2650,
  "University D",   2011,      5500000,    550000,         2300000,
1471,             648,
  "University D",   2013,      6350000,    650000,         3150000,
1690,             746,
  "University D",   2015,      7000000,    700000,         3100000,
1828,             851,
  "University D",   2017,      8050000,    750000,         3200000,
2024,             891,
  "University D",   2019,      9000000,   1100000,         3700000,
2300,             918
  )
```

We first proceed as we did in the previous chapter. We adjust for the real value of money by converting all dollars to 2011 dollars. Then we create our new adjusted publications variable by assigning extra value to the most highly cited articles. Our method, as described above, is to simply count the highly cited publications twice.

```
# First get put the new data together
rvm_2007 <- tibble(
  year = c(2011, 2013, 2015, 2017, 2019),
  rvm = c(1.000, 0.952, 0.896, 0.850, 0.812)
)


# adjust all monetary values for each institution to 2007 dollars
unis_adjusted_res <- uni_data_res %>%
  group_by(institution) %>%
  mutate_at(vars(starts_with("x")), # performs function below on
each column beginning with "x_"
            function(col) col*rvm_2007$rvm) %>%
  ungroup() %>%
  mutate(y_adjusted_pubs = y_pubs_total + y_pubs_top25)


glimpse(unis_adjusted_res)
## Rows: 20
## Columns: 8
## $ institution        <chr> "University A", "University A",
"University A", "Un...
## $ year          <dbl> 2011, 2013, 2015, 2017, 2019, 2011, 2013,
2015, 201...
## $ x_salaries    <dbl> 18240000, 18564000, 22256640, 24735000,
24408720, 6...
## $ x_cap_ex       <dbl> 5640000, 6854400, 7150080, 9333000,
9159360, 105000...
```

```
## $ x_intermediate  <dbl> 8760000, 10281600, 11934720, 13107000,
13300560, 40...
## $ y_pubs_total    <dbl> 3885, 4728, 6027, 7297, 8083, 3025,
3336, 3803, 405...
## $ y_pubs_top25    <dbl> 1615, 2073, 2770, 3615, 4020, 1253,
1460, 1744, 176...
## $ y_adjusted_pubs <dbl> 5500, 6801, 8797, 10912, 12103, 4278,
4796, 5547, 5...
```

Our next step is to combine this new information with the data from Chap. 4. In the USA, universities report education and research expenditures separately. Not all countries' higher education authorities require such accounting methods. To keep the content most relevant for broader audiences, we combine the education and research expenditures together and make the assumption of joint production processes as described above. If you are keen to explore separate productivity of education and research, you can apply the method from the previous chapter to the data we've just prepared.

```
# add research expenses to education expenses, and
# include research output variables
unis_adjusted <- unis_adjusted_edu %>%
  mutate(x_salaries = x_salaries + unis_adjusted_res$x_salaries,
         x_cap_ex = x_cap_ex + unis_adjusted_res$x_cap_ex,
           x_intermediate = x_intermediate + unis_adjusted_res$x_
intermediate) %>%
  bind_cols(select(unis_adjusted_res, y_pubs_total, y_pubs_top25,
y_adjusted_pubs))

# check out the new data
glimpse(unis_adjusted)
## Rows: 20
## Columns: 11
## $ institution        <chr> "University A", "University A",
"University A", "Un...
## $ year            <dbl> 2011, 2013, 2015, 2017, 2019, 2011, 2013,
2015, 201...
## $ x_salaries      <dbl> 30400000, 30940000, 37094400, 41225000,
40681200, 1...
## $ x_cap_ex        <dbl> 9400000, 11424000, 11916800, 15555000,
15265600, 21...
## $ x_intermediate  <dbl> 14600000, 17136000, 19891200, 21845000,
22167600, 8...
## $ y_fte_course    <dbl> 20900, 23640, 25580, 26980, 29600, 9830,
12080, 135...
```

```
## $ y_compl_course   <dbl> 3940, 4554, 6414, 6224, 6444, 2122,
1737, 2680, 358...
## $ y_adjusted_load  <dbl> 24840, 28194, 31994, 33204, 36044,
11952, 13817, 16...
## $ y_pubs_total     <dbl> 3885, 4728, 6027, 7297, 8083, 3025,
3336, 3803, 405...
## $ y_pubs_top25     <dbl> 1615, 2073, 2770, 3615, 4020, 1253,
1460, 1744, 176...
## $ y_adjusted_pubs  <dbl> 5500, 6801, 8797, 10912, 12103, 4278,
4796, 5547, 5...
```

The data now represent all the institutional information we need for this tutorial. We'll use the data again later in Chap. 8 too, so let's save now. Also, let's get our data into long format for easier processing.

5.4.4 Examine Institutional Change Profiles

We now have a great deal of data to explore. So let's start to wrap our heads around some of the key factors that may be driving broader trends. Our objective for this section of the tutorial is to understand how each of our input-output components has changed over the period to get an idea for what our productivity results may end up looking like. We do this not only for our own initial understanding but also because if it becomes necessary later to justify or explain final results, we'll already have built a nuanced understanding of the data.

Our first step is to use the calc_change() function we created in Chap. 4 to find the change index and cumulative change values from unis_adjusted_long. After that, we'll flag each variable that will be used directly in the productivity calculation. We create the flag for two reasons. First, it will make filtering the data easy once we are ready to calculate productivity change indicators. Second, it will allow for some quick comparisons in the analysis that follows directly.

```
#run function
uni_elements_change <- unis_adjusted_long %>%
  calc_change(var = "value",
              calc = "change_index",
              institution, variable) %>%
  calc_change(var = "value_index",
              calc = "cum_change",
              institution, variable) %>%

  # add a categorical column to help us distinguish between
variables
  # we'll be using directly in the productivity calculation and the
  # raw component data we used to create our adjusted load and
```

```
mutate(element_usage = case_when(
    grepl("adjusted", variable)              ~ "Key productivity
component",
    nature == "x"                     ~ "Key productivity component",
    TRUE                              ~ "Raw output component"
))

glimpse(uni_elements_change)
## Rows: 180
## Columns: 8
## $ institution        <chr> "University A", "University A",
"University A", "U...
## $ variable           <chr> "x_cap_ex", "x_cap_ex", "x_cap_ex",
"x_cap_ex", "x...
## $ nature             <chr> "x", "x", "x", "x", "x", "x", "x", "x",
"x", "x", ...
## $ year               <dbl> 2011, 2013, 2015, 2017, 2019, 2011,
2013, 2015, 20...
## $ value              <dbl> 9400000, 11424000, 11916800, 15555000,
15265600, 1...
## $ value_index        <dbl> 1.000000, 1.215319, 1.043137, 1.305300,
0.981395, ...
## $ cum_change_value <dbl> 1.000000, 1.215319, 1.267745, 1.654787,
1.624000, ...
## $ element_usage      <chr> "Key productivity component", "Key
productivity co...
```

Let's now see the cumulative change in each data element over the period by filtering only for the year 2019 and plotting the change for each variable. Using the new element_usage variable, we'll highlight the variables to be used directly in the productivity calculation (Fig. 5.2).

```
# format data for plotting
plot_data <- uni_elements_change %>%
    filter(year == 2019) %>%
    mutate(element_usage = ifelse(element_usage == "Key productivity
component",
                                "Key productivity \ncomponent",
                                "Raw output \ncomponent"))
# plot
plot_data %>%
    ggplot(aes(x = variable, y = cum_change_value, alpha = element_
usage)) +
    geom_col() +
    facet_wrap(vars(institution)) +
```

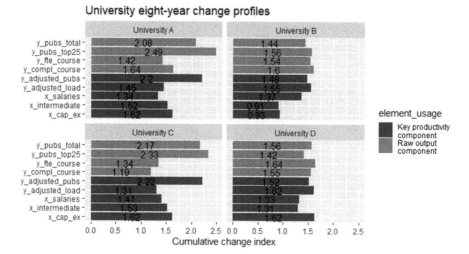

Fig. 5.2 Change scores for all university data elements

```
scale_alpha_manual(values = c(1, .6)) +
  geom_text(aes(label = round(cum_change_value, 2)), position =
position_stack(vjust = 0.5), alpha = 1) +
  labs(y = "Cumulative change index",
      x = "",
      title = "University eight-year change profiles",
      fill = "Element \nusage") +
coord_flip()
```

The 8-year change profiles show the areas where the university has been growing fastest and slowest in relation to other variables and other institutions. Where we see trends of faster output growth than input growth, we can infer that productivity change will be positive. We can also see which elements are driving the change. Further, for our adjusted load and adjusted completions variables, we can see whether quality or quantity is driving change. For example, University A has seen strong growth in adjusted publications, and their quick rise in highly cited articles is most behind that trend.

5.4.5   Calculate Productivity

We now further optimize the code from Chap. 4 for calculating productivity change. We need a simpler calculation for our productivity threshold analysis. If we are going to produce a range of possible productivity portrayals, we need a single function to work with. The new productivity change calculation function will take five arguments. The first argument, as always, will be the data set the function is operating on. The second and third arguments will be the respective weights that we'll assign to the education and research outputs. We'll give these weights a default value of 50-50. Without a clear rationale, there is no reason to give education or research a relatively higher value weight by default. The fourth and fifth arguments will serve to specify which output is the education output and which is the research output.

```r
# specify function arguments
prod_change_calc <- function(df,
                             edu_weight = 0.5,
                             res_weight = 0.5,
                             edu_var = "y_adjusted_load",
                             res_var = "y_adjusted_pubs") {

  # prep data
  prod_data_long <- df %>%
    filter(element_usage == "Key productivity component") %>%
    select(-element_usage) %>%
    group_by(institution, year, nature) %>%
    mutate(prop = ifelse(nature == "x", value/sum(value), NA)) %>%
    ungroup() %>%
    group_by(institution, variable) %>%
    mutate(weight = case_when(
      nature == "x"            ~ (prop + lag(prop))/2,
      variable == edu_var      ~ edu_weight,
      variable == res_var      ~ res_weight
      )) %>%
    ungroup()

  # calculate productivity change
  prod_change <- prod_data_long %>%
    mutate(weighted_index = value_index^weight) %>%
    group_by(institution, nature, year) %>%
    summarise(composite_index = prod(weighted_index)) %>%
    ungroup()  %>%
    pivot_wider(names_from = "nature", values_from = "composite_
index") %>%
    mutate(P = y/x) %>%
    pivot_longer(x:P, names_to = "nature", values_to = "compos-
ite_index") %>%
    arrange(institution, nature, year) %>%
    calc_change(var = "composite_index",
                calc = "cum_change",
                institution, nature) %>%
    mutate(nature = ifelse(nature == "P",
```

```
                                                           glue ("{nature}:
e{edu_weight*100}-r{res_weight*100}"),
                                      nature))

    return (prod_change)
}


# save for use later
save (prod_change_calc, file = "data/prod_change_calc.rda")
```

Let's now assume that education and research outputs have roughly the same value in aggregate.

```
# run function
prod_change_res_edu <- prod_change_calc(df = uni_elements_change)

# view result
glimpse (prod_change_res_edu)
## Rows: 60
## Columns: 5
## $ institution          <chr> "University A", "University A",
"University A"...
## $ year                 <dbl> 2011, 2013, 2015, 2017, 2019, 2011,
2013, 2015...
## $ nature               <chr> "P: e50-r50", "P: e50-r50", "P:
e50-r50", "P: ...
## $ composite_index      <dbl> 1.0000000, 1.0831424, 1.0462276,
0.9942696, 1....
## $ cum_change_composite <dbl> 1.000000, 1.083142, 1.133213,
1.126720, 1.2444...
```

We can see that we've now got a new type of observation for the nature column called "P: e50-r50" signifying productivity change with equal emphasis on education and research. Let's now plot this data using the plot_prod_change() function we created in the previous chapter to compare composite input and composite output growth with the new productivity indicator (Fig. 5.3).

```
plot_prod_change(prod_change_res_edu)
```

Examining the figures, we notice that in each case the rate of increase of outputs exceeds the rate of increase of inputs. We thus don't see any negative productivity change. Recalling these universities' productivity change results from Chap. 4, factoring research into the equation appears to have improved their performance outlook. We can better understand the education-research dynamic however by proceeding to threshold analysis.

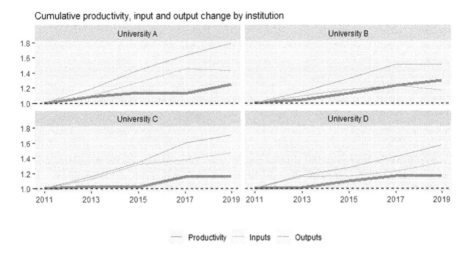

Fig. 5.3 Cumulative productivity change with equal weight on research and education

5.4.6   Threshold Analysis

Threshold analysis will generate a range of plausible productivity portrayals using the edu_weight and res_weight arguments from the prod_change_calc() function we created above. The range of productivity portrayals helps us determine the extent to which a favorable portrayal depends upon positive performance in one domain or the other.

First create a range of corresponding values for our education and research weights. Recall that as per the definition of the Törnqvist Index (TI), the respective value shares of education and research must sum to 1 or 100%, representing the total value produced by the institution in these domains. Place an upper and lower bound on each of the values at 0.7 and 0.3, respectively. These bounds represent the assumption that, of the total output value produced by an institution, neither education nor research outputs by themselves would ever exceed 70% of that value, or conversely, less than 30%. Thus, we define a sufficient set of combinations of these values that covers the 70-30 range.

```
# create res and edu value weight vectors
edu_vals <- seq(from = 0.3, to = 0.7, by = 0.05)
res_vals <- seq(from = 0.7, to = 0.3, by = -0.05)

# view the weight ranges as a table
tibble(edu_vals, res_vals)
## # A tibble: 9 x 2
##    edu_vals res_vals
##       <dbl>    <dbl>
## 1      0.3      0.7
```

```
## 2        0.35      0.650
## 3        0.4       0.6
## 4        0.45      0.550
## 5        0.5       0.500
## 6        0.55      0.450
## 7        0.6       0.400
## 8        0.65      0.350
## 9        0.7       0.3
```

The next step is to run our prod_change_calc() on our uni_elements_change data set for each combination of the education and research value weights above. The method we choose to do this comes from the purrr package, which is automatically activated as part of the tidyverse. The function we'll use to iterate prod_change_calc() over each combination comes from the map() family of functions. We'll use map2_dfr(). The "2" signifies that we'll be iterating over two sets of information, edu_vals and res_vals. The "dfr" means that we want the output to be a data frame, such that each subsequent iteration of the function is bound by rows to the previous iteration. For example, when we ran the prod_change_calc() above, we were given a data frame containing 5 columns and 60 rows. We know that we're going to be iterating that function over nine unique combinations of education and research value weights. Thus, the output of using map2_dfr() over that function and these sets of values will produce a single data frame containing five columns, but will now contain 60*9 = 540 rows.

The arguments for the map() family of functions go as follows. The first argument(s) will be the set(s) of values used for iteration, which must be arguments of the function that we'll be iterating. The next map() arguments must be the function call itself (prod_change_calc). Any additional arguments are optional, but must also be arguments of the function that's being iterated. In our case, the additional argument will be df, containing all the data element change values that will be aggregated into productivity change scores using the different combinations of education and research value weights.

```
# map the function over each combination of res and edu value weights
prod_range <- map2_dfr(edu_vals, res_vals,
                       prod_change_calc,
                       df = uni_elements_change)
# check the dimensions
dim(prod_range)
## [1] 540     5
# save for use later
save(prod_range, file = "data/prod_range.rda")
```

We can see now that we have created a very long data frame, resulting from having iterated the prod_change_calc() function nine times. And we can take a further glimpse at the data to prove that the map() function worked.

```
glimpse(prod_range)
## Rows: 540
## Columns: 5
## $ institution          <chr> "University A", "University A",
"University A"...
## $ year                 <dbl> 2011, 2013, 2015, 2017, 2019, 2011,
2013, 2015...
## $ nature               <chr> "P: e30-r70", "P: e30-r70", "P:
e30-r70", "P: ...
## $ composite_index      <dbl> 1.0000000, 1.1018604, 1.0739799,
1.0303713, 1....
## $ cum_change_composite <dbl> 1.000000, 1.101860, 1.183376,
1.219317, 1.3524...
```

Now we check what all the productivity results look like and visualize the productivity thresholds. Let's define a plotting function to accomplish this because
we'll use it again later in Chap. 8 (Fig. 5.4).

```
# visualise results
plot_prod_range <- function(df, var) {

  plot <- df %>%
    filter(!nature %in% c("x", "y")) %>%
    ggplot(aes(x = year, y = .data[[var]], color = nature)) +
    facet_wrap(~institution) +
    geom_line() +
    geom_hline(yintercept = 1, linetype = "dashed") +
```

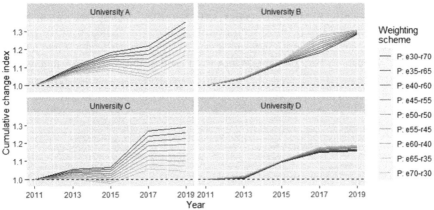

Fig. 5.4 Visualising productivity threshold analysis for four different institutions

```
    scale_color_grey() +
    scale_x_continuous(breaks = c(2011, 2013, 2015, 2017, 2019)) +
    labs(x = "Year",
         y = "Cumulative change index",
         color = "Weighting \nscheme",
            title = "Range of plausible productivity scores with \
nvariable education and research value weights")

  return(plot)
}


save(plot_prod_range, file = "data/plot_prod_range.rda")


plot_prod_range(prod_range, var = "cum_change_composite")
```

First notice the difference in the overall performance profiles of these institutions. University B and D's profiles are more robust to changes in the value weights than Universities A and C. Further, both B and D achieve more positive productivity change scores when education outputs are weighted most heavily. Universities A and C, however, achieve their most favorable productivity scores when research is weighted most heavily. Interpreting the results further takes significant care. We'll consider different uses and interpretations of the data in the final section below and reflect on use of this technique in practice.

Before we move to reflections, however, let's examine our new data from one additional angle—performance rankings. The example below illustrates how sensitive most institutions are to different ranking systems.

5.4.7 Rankings Analysis

It's common to want to rank universities. Let's see what happens when we examine the institutions above in terms of their overall productivity change over the period. We'll repeat a similar exercise in Chap. 8, so let's wrap up the analysis into a function. We'll create three separate rankings from each of the following value weight schemes: "P: e30-r70," "P: e50-r50," and "P: e70-r30." These schemes represent the upper and lower bounds, as well as equal weights for education and research.

```
# define function parameters
get_ranks <- function(df,
                       var,
                       yr = 2019,
                       rank_weights = c("P: e30-r70", "P: e50-r50",
"P: e70-r30")) {
```

```
# filter and select key variables
prod_rankings <- df %>%
   filter(year == yr,
          nature %in% rank_weights) %>%
   select(institution, year, nature, .data[[var]]) %>%
   arrange(nature, desc(.data[[var]])) %>%

   # create rankings
   group_by(nature) %>%
   mutate(rank = 1:n()) %>%
   ungroup() %>%
   select(-.data[[var]]) %>%

   # pivot for easier readability
   pivot_wider(names_from = "nature", values_from = "rank")

   return(prod_rankings)
}

# save for later
save(get_ranks, file = "data/get_ranks.rda")
```

Run the function to get rankings (Table 5.1).

```
# run function
prod_rankings          <-        get_ranks(prod_range,        var       =
"cum_change_composite")

# check results
prod_rankings
```

Notice how sensitive each university is to the different value weighting schemes. Depending on ranking criteria, each institution could find itself either at the top of the field or at the bottom of the field. The example demonstrates not only how malleable higher education institutional data is to manipulation, but also it shows that when the results alternative ranking criteria are not shown side by side, very little can be learned from the numbers on the league tables. To engage further with the parameters of this threshold analysis please vist ikenresearch.com/interact.

Table 5.1 Alternative rankings based on different criteria

Institution	Year	P: e30-r70	P: e50-r50	P: e70-r30
University A	2019	1	2	3
University C	2019	2	4	4
University B	2019	3	1	1
University D	2019	4	3	2

5.5   Reflections on Threshold Analysis in Practice

Using value weight ranges to generate productivity thresholds adds another dimension to productivity change analysis. The range of portrayals for each institution produces insight that would otherwise stay hidden under a single treatment of the data. When reasonable upper and lower bounds are placed on performance, key institutional strengths can be revealed, and they open the door for more contextual explanation and investigation.

For example, the smaller threshold of University D and the larger threshold of University B indicate nothing "good" nor "bad" in themselves. Rather, they inspire more complex questions about whether observed performance aligns with policies and priorities. University B's most favorable productivity portrayal depends on weighting research outputs more heavily. Does this reflect where the university has been investing its resources? Do trends align with where the institution wants to go, or are education priorities being left behind? For University D, bot education and research productivity are climbing in lock-step, but what was the baseline, and what was the objective? Would University D like to have seen productivity gains in education outpace research after a large course consolidation effort?

The thresholds highlight the importance of not only entertaining multiple performance scenarios, but also they also signpost where further analysis should be focused. The thresholds reveal why analysis must be collaborative and why exercising judgment and accounting for context can never be excluded from a rigorous quantitative analysis on higher education institutions.

The rankings exercise is of course extra sensitive because only four institutions are included. The message here, though, is that it's not difficult to engineer complex institutional data to suit individual purposes. While the very top universities in the world are often quite robust to different weightings of various indicators over time, the bottom 95% of institutions are very diverse with unique strengths and weaknesses. A single ranking method does very little to communicate overall performance across a field of institutions.

A final note on this exercise is that universities accomplish more than just graduating students and accumulating affiliated research publications. More comprehensive performance assessments would incorporate more qualitative indicators such as international and disciplinary collaborations, industry engagement, and community service. The indicators we have chosen for this analysis happen to be ones that can be compared across almost any field of institutions. They further represent outputs of the essential academic promises of universities—creating and disseminating knowledge. And these activities are the ones to which universities are held to account during planning and budgeting activities. More bespoke analyses are possible, and they should be conducted. As long as outputs are observable and objective, these methods can and should be adapted to examine broader spectrums of performance and productivity.

Frontier Analysis with DEA

6

6.1 Objective

This chapter introduces productivity frontier analysis and explains the detail behind a common technique for conducting frontier analysis, called data envelopment analysis (DEA). By the end of the chapter, the reader will not only understand how the DEA algorithm works but also will be able to construct an investigation using DEA from the ground up with a linear program. The chapter first builds a conceptual understanding of what DEA is and why it has become popular for measuring productivity in education institutions and for NFPs. The chapter demonstrates the power and limitations of DEA.

6.2 What Is Frontier Analysis?

Frontier analysis uses empirical methods to estimate productivity thresholds. It differs from the threshold analysis introduced in Chap. 5 because frontier analysis determines thresholds from characteristics of the dataset, rather than from judgments made by the analyst. Frontier analysis is used for benchmarking the maximum achievable, feasible level of productivity that any member of a field of peer organizations should be able to achieve. Frontier analysis establishes a frame of reference for the upper bound on performance excellence, called the production frontier. The production frontier allows the analyst to quantify concrete gaps in performance relative to the frontier and between peer organizations.

A common term used to describe the productivity of an organization relative to the frontier is "technical efficiency" (TE). Estimating TE relies on an assumption that all organizations in the data set have access to the same level of technology or innovative processes to generate their outputs. TE is a normalized productivity score, where the most efficient organization in the field is assigned a score of 1.0

© The Author(s), under exclusive license to Springer Nature
Switzerland AG 2021
K. Moore, *Measuring Productivity in Education and Not-for-Profits*,
Management for Professionals, https://doi.org/10.1007/978-3-030-72965-3_6

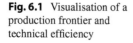

Fig. 6.1 Visualisation of a production frontier and technical efficiency

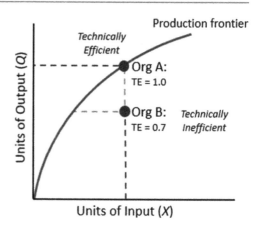

(representing the frontier). We used the same principle in Chap. 3. Figure 6.1 shows one organization achieving frontier-level efficiency and another organization at a TE of 70%.

DEA is one of the most popular methods for frontier analysis and productivity analysis in general in education and not-for-profit organizations, including healthcare institutions. The purpose of this chapter is to give analysts the power to understand and apply the results of DEA to their fullest extent. One reason for the technique's popularity is that DEA provides a mathematical solution for finding productivity scores when organizations have multiple, non-priced inputs and outputs. In other words, the DEA algorithm finds an empirically justified set of value weights for non-priced elements in the dataset. A full threshold analysis like the one we performed in Chap. 5 thus becomes unnecessary. With reference to the higher education scenario from the previous chapter, the DEA algorithm would comb through the data several times and find an empirically justified set of values for the education and research value weights for each institution. The benefit of DEA is that results are formed on an objective basis, reducing human bias.

Reduction of human bias is also the drawback of the technique. As discussed in the reflection section of Chap. 5, university productivity estimates must be treated with care if results are intended to inform policy, decision-making, or true bench-marking. DEA has significant power and utility, but any amount of human bias it removes from the equation, it replaces with computer bias—and the DEA algorithm is an order of magnitude more complex than TI, so appropriate interpretation of results is more difficult. Analysts often do not perform thorough investigations of the assumptions underpinning DEA calculations—along with their implications—and results in the literatyre are frequently presented as definitive.

6.3   A Detailed Look at Frontier Analysis with DEA

6.3.1   A Not-for-Profit Example

As we proceed through this chapter, we assume a common interest in understanding how DEA works. This may, however, be your introduction to the technique, so we'll first proceed with an example to ease into the technical detail. In contrast to the preceding chapters, we'll introduce our scenario up front and use it to help explain key concepts and applications of the DEA technique.

Consider the following scenario. A government agency has approved a new program which establishes a grant to contract different services providers in different cities to deliver training workshops for health workers who need certifications. The agency wants to conduct a formative, process evaluation of the program halfway through the grant period. The providers' quarterly reports to the agency suggest that four indicators should be assessed—two inputs and two outputs—to determine the productivity of each organization. The inputs are the size of the training facilities used in each city and the number of full-time staff organizing and conducting the workshops. The outputs are the number of participants who attended each workshop and also the number of those participants who successfully completed the trainings to earn a certification. As the evaluation is only processes-focused, the agency is not yet concerned with anything resembling a financial audit. Rather, the agency wants an operational analysis, the results of which can be shared, so each organization can strive for best practice during the second half of the grant period.

In order to determine each organization's productivity, we must find a way, without prices, to aggregate square feet with people and aggregate people with certificates. Similar to our approach in Chap. 5 with non-priced outputs, we thus need to find some representation of the value of these data elements, so we can aggregate them meaningfully into a single productivity ratio. The framework for doing so is shown below, where u(i) and v(i) represent the unit values of the outputs and inputs, respectively.

$$\text{productivity} = \frac{u_1\left(number\ of\ workshops\right) + u_2\left(number\ of\ certifications\right)}{v_1\left(square\ feet\right) + v_2\left(number\ of\ staff\right)}$$

DEA is the tool we'll use to find each u(i) and v(i). How then does the DEA algorithm come up with these values in a fair, realistic, consistent, and empirical way? The DEA algorithms are designed to assign numerical values to u1, u2, v1, v2, such that the organization is portrayed in its most favorable light with respect to the other organizations. Without any upper or lower bounds placed on the potential values of each u(i) and v(i), DEA finds these values for each organization, such that they appear as productivity as possible. One prime assumption here is that organizations will tend to play to their strengths, so where the organization is performing well, performance in that domain will be emphasized. DEA thus judges each organization according to its strengths, and the organization that is leveraging its strengths most effectively will rise to the top. From here, we can best illustrate this idea by adding some data.

6.3.2   DEA Objective

First consider the organizations in the table below. Also notice each one is listed in a column, named "dmu" or decision-making unit (DMU). DMU is the term most commonly used for the organizations or institutions being considered for a DEA analysis (Table 6.1).

If this data were loaded into DEA program, it would answer the following question: for each organization, how can we maximize its productivity ratio? For example, take Organization A. What values of u1, u2, v1, v2 produce the highest possible score for Organization A when the other organizations in the data set are given those same value weights? And subsequently, what are the unique value weighting schemes for Organization B, C, D, and E that accomplish the same thing?

In order to answer these questions, we need to solve an optimization problem for each individual organization. That is, what DEA does. The optimization algorithm is both the beauty and the pitfall of DEA. On one hand, DEA prevents unfair representations of the organizations because the value weights given to each one numerically produce the most favorable portrayal possible. On the other hand, determining whether the value weights are truly founded in reality is difficult. Whether the weighting scheme produces productivity indicators that are representative of true value or authentic contextual circumstances is not clear.

6.3.3   DEA Algorithm Walk-Through

So let's look at how this works in practice. We use Organization A as an example, with the understanding that DEA repeats these steps for the other four organizations. We use the data from above.

Objective

$$Maximize\frac{u_1 20 + u_2 200}{v_1 600 + v_2 4}$$

Subject to the constraints:

$$\frac{u_1 20 + u_2 250}{v_1 800 + v_2 5} \leq 1 \ (Org B constraint)$$

Table 6.1 Example data for DEA

dmu	x_sqr_ft	x_staff	y_participants	y_certifications
Org_A	600	5	200	100
Org_B	800	5	250	125
Org_C	800	6	300	125
Org_D	300	4	175	140
Org_E	350	4	200	155

$$\frac{u_1 20 + u_2 300}{v_1 1000 + v_2 6} \leq 1 \; \left(\text{OrgC constraint}\right)$$

$$\frac{u_1 24 + u_2 200}{v_1 800 + v_2 4} \leq 1 \; \left(\text{OrgD constraint}\right)$$

$$\frac{u_1 30 + u_2 300}{v_1 1000 + v_2 6} \leq 1 \; \left(\text{OrgE constraint}\right)$$

$$u_1, u_2, v_1, v_2 \geq 0 \left(\text{non} - \text{negative value constraints}\right)$$

The question you may (and should) ask here is, "why are all the constraints set to less than or equal to one?" The answer is because setting this constraint accomplishes the same normalizing effect as was used in Chap. 3 with the school districts example. Productivity is set to a maximum value of 1, or 100%, representing the production frontier. Notice also that there is a non-zero constraint as well, meaning that productivity scores can range only from one to zero. If an organization emerges from the algorithm with a score of 0.9, then we interpret the result as 90% productive relative to the maximum. The DEA algorithm essentially has an in-built feature-scaling of results, oriented around the organization that emerges with the highest score.

The DEA algorithm is an application of linear programming. Notice, however, that the objective and constraints above are not linear formulas in the form of $y = mx + b$. Rather they are rational expressions because of the productivity ratio we are trying to optimize. There is a clever work-around, however. We need to restate the formulae in a linear format, preserving their meaning.[1]

There is more than one way to construct the linear program, however, and each has implications for the formulation of parameters and the interpretation of results. We now must ask ourselves an essential question about how these organizations are likely to think about and pursue productivity improvements:

> Are these organizations more likely to pursue improved productivity through better utilization of a fixed amount of inputs, or are the likely more likely to set standards for service levels and then use as few resources as possible to meet that standard?

That is, given a fixed set of resources, do they want to maximize value for their communities? Or do they have a fixed set of responsibilities and want to minimize costs and consumption? The answer has both practical and operational consequences for the organizations as well as modeling implications for the DEA algorithm.

Let's assume these organizations have a fixed budget and are trying to best utilize their resources to maximize output. This is often the case with NFPs and education institutions. We can now adjust our objective function above to reflect this goal: maximize the numerator (output) for a fixed denominator (input). We now have a simpler objective function. This is called an "output-oriented" approach.

---

[1] Charnes, A., Cooper, W. W., & Rhodes, E. (1978). Measuring the efficiency of decision making units. European journal of operational research, 2(6), 429-444.

Objective

$$Maximize\, u_1\, 20 + u_2\, 200$$

The other organizational constraints must also be adjusted. Consider the following. If the value of a fraction is less than or equal to 1, then the numerator must be less than or equal to the denominator. Hence, if you subtract the denominator from the numerator, you must either get a negative number or zero. In other words,

$$\frac{a}{b} \leq 1 \quad \text{implies} \quad a - b \leq 0$$

From this, we can also restate the constraints to construct well-formed linear program.

Constraints

$$u_1\, 20 + u_2\, 250 - \left(v_1\, 800 + v_2\, 5 \right) \leq 0 \quad \left(\text{Org B constraint} \right)$$

$$u_1\, 20 + u_2\, 300 - \left(v_1\, 1000 + v_2\, 6 \right) \leq 0 \left(\text{Org C constraint} \right)$$

$$u_1\, 24 + u_2\, 200 - \left(v_1\, 800 + v_2\, 4 \right) \leq 0 \quad \left(\text{Org D constraint} \right)$$

$$u_1\, 30 + u_2\, 300 - \left(v_1\, 1000 + v_2\, 6 \right) \leq 0 \left(\text{Org E constraint} \right)$$

$$v_1\, 600 + v_2\, 4 = 1 \left(\text{fixed input constraint} \right)$$

$$u_1, u_2, v_1, v_2 \geq 0 \left(\text{non} - \text{negative value constraints} \right)$$

6.4   Tutorial: Which NFP Is the Top Performer?

6.4.1   Setup

First, to run the code in this tutorial, you'll need the packages below.

```
library(tidyverse)
library(knitr)
library(lpSolve)  # Linear programming package
library(rDEA)     # Automatic DEA functions
```

6.4.2   Introduction

We begin the tutorial with a data set similar to the sample above. And we add one extra column x_workshops and an extra organization, "Org_F," to increase the complexity and to demonstrate the power of the DEA technique. In this scenario, workshops delivered are considered an input because of all the materials and expenses associated with workshop delivery. Assume that each organization is contracted to deliver a fixed number of workshops and the aim is to upskill as many participants as possible.

6.4.3 Create New Data Set

Notice the standard wide format for inputs and outputs. The primary reason for using wide format instead of long format in this tutorial is because data in this form is more suitable for constructing a linear program and using the lpsolve package.

```
dea_data_01 <- tribble(
    ~dmu,    ~x_sqrft,    ~x_staff,    ~x_workshops,    ~y_participants,
  ~y_certs,
    "Org_A",      600,        6,           20,              195,        105,
    "Org_B",      775,        6,           20,              240,        132,
    "Org_C",      800,        7,           20,              285,        131,
    "Org_D",      300,        5,           35,              175,        145,
    "Org_E",      350,        5,           40,              195,        155,
    "Org_F",      400,        5,           40,              205,        155)

# save for use in next chapter
save(dea_data_01, file = "data/dea_data_01.rda")
```

For full clarity, the inputs denote the size of the training facility in square feet, the number of full-time staff, and the number of workshops delivered, respectively. The outputs denote the number of participants who participated in the workshops and the number of those participants who earned certifications.

Next, we save some of the features of this data set as objects for use in the upcoming analysis.

```
# number of dmus
N <- nrow(dea_data_01)

# dmu names
dmu_names <- dea_data_01$dmu

# isolate inputs
inputs <- select(dea_data_01, starts_with("x"))

# isolate outputs
outputs <- select(dea_data_01, starts_with("y"))

# number of input variables
```

```
x_dim <- ncol(inputs)

# number of output variables
y_dim <- ncol(outputs)

# variable names (order outputs then inputs)
var_names <- c(colnames(outputs), colnames(inputs))
```

6.4.4   Define a Linear Program for Org A

We can now define a linear program that corresponds to the variables above. We'll start with a linear program designed to find value weights for Org A's inputs and outputs, which maximize its productivity ratio. The steps go as follows. First, we define our objective function. Second we'll create a matrix that represents all the information that falls on the left-hand side of the constraint equations, as in the *figure above*. Third, we create a vector of constraint values, which corresponds to all the values that fall on the right-hand side of the constraint equations. Fourth, we create a final vector to define all the inequality and equivalence relationships between the left- and right-hand sides of the constraint equations. Finally, we input all that information into arguments of the lp() function from the lpSolve package to solve the linear program.

Create the objective function.

```
# objective function
objective <- c(as.numeric(outputs[1,]), rep(0, x_dim))

# view result
print(objective)
## [1] 195 105 0 0 0
```

Recall that since we are taking an output-oriented approach, our only objective is to maximize the outputs for Org A, which we've listed first. There are a total of five variables in this linear program, however, so we always have to account for them, hence the three zeros still present in the vector. The lp() function uses matrix algebra; we have to keep a consistent structure for each element we create as an argument to that function. The zeros are just placeholders to let the program know that there are three other variables that we'll be using as constraints later, but their values will not directly affect the maximum productivity calculation that we're after.

Next, we reshape the input-output data to fit the form of the constraint equations shown in the *figure above*. Remember that all inputs are being subtracted from the outputs in that equation, so we need to assign negative values to all the inputs. We call this new object the

"auxiliary matrix."

```
# create output oriented aux matrix for constraints
aux <- cbind(as.matrix(outputs), as.matrix(-1*inputs))

print(aux)
## y_participants y_certs x_sqrft x_staff x_workshops
## [1,]            195     105    -600     -6          -20
## [2,]            240     132    -775     -6          -20
## [3,]            285     131    -800     -7          -20
## [4,]            175     145    -300     -5          -35
## [5,]            195     155    -350     -5          -40
## [6,]            205     155    -400     -5          -40
```

We now combine the auxiliary matrix with all the other non-negativity and input constraints noted in the *figure above*.

```
# First, select the constraints for all Orgs except Org A
var_constraints <- rbind(aux[-1,],

                         # assign non-negativity constrains for
each variable
                         diag(x_dim+y_dim),

                         # hold inputs constant for the DMU we're
optimising
                         c(rep(0, y_dim), as.matrix(inputs[1,])))
)

print(var_constraints)
##        y_participants y_certs x_sqrft x_staff x_workshops
## [1,]             240     132    -775     -6          -20
## [2,]             285     131    -800     -7          -20
## [3,]             175     145    -300     -5          -35
## [4,]             195     155    -350     -5          -40
## [5,]             205     155    -400     -5          -40
## [6,]               1       0       0      0            0
## [7,]               0       1       0      0            0
## [8,]               0       0       1      0            0
## [9,]               0       0       0      1            0
## [10,]              0       0       0      0            1
## [11,]              0       0     600      6           20
```

We now set the constraint values, i.e., the right-hand side of the constraints equation. The first five constraints must be set to less than or equal to zero. We then specify that each value weight we want to find must be greater than or equal to zero, and finally, we want to hold Org A's inputs constant, so we set them equal to 1.

```
# right hand side constraints
```

```
rhs_constraints <- c(rep(0, N-1), rep(0, x_dim + y_dim), 1)

print(rhs_constraints)
## [1] 0 0 0 0 0 0 0 0 0 0 1
```

Now we create a corresponding vector of all the equality and inequality relationships we just mentioned above.

```
# constraint directionality
const_dir <- c(rep("<=", N-1), rep(">=", x_dim + y_dim), "=")

print(const_dir)
## [1] "<=" "<=" "<=" "<=" "<=" ">=" ">=" ">=" ">=" ">=" "="
```

We're now ready to run the optimization analysis for Org A (Table 6.2).

```
# run linear program
dmu_optimum <- lp(direction = "max",
                  objective.in = objective,
                  const.mat = var_constraints,
                  const.dir = const_dir,
                  const.rhs = rhs_constraints)

# Organise the outputs of the linear program into a results table
results <- tibble("dmu" = rep(dmu_names[1], 1 + x_dim + y_dim),
             "variable" = c("Optimal Efficiency", paste("Weight",
var_names)),
                          "value"  =  c(dmu_optimum$objval,
dmu_optimum$solution)
)

# view results
results
```

Table 6.2 Optimization results for Org A

dmu	Variable	Value
Org_A	Optimal efficiency	0.9016999
Org_A	Weight y_participants	0.0005814
Org_A	Weight y_certs	0.0075079
Org_A	Weight x_sqrft	0.0007461
Org_A	Weight x_staff	0.0000000
Org_A	Weight x_workshops	0.0276156

There is much to gather from these results. First, we can see that in the best case scenario for Organization A, its productivity level is about 90% of frontier efficiency for the field of organizations—or 90% as productive as the most productive organization in the field. We can also see which data elements are emphasized or de-emphasized across the field of organizations for Org A to achieve its optimal portrayal.

Consider the output weights as an example. y_certs is weighted more heavily than y_participants. This may imply one of two situations. Either Org A is performing relatively well in terms of the number of certifications it's awarding, and so the algorithm has chosen to emphasize this aspect of performance for Org A. Or conversely, the other organizations are performing relatively better in terms of the amount of participants they are attracting, and so the algorithm has chosen to downplay this aspect of performance for Org A's benefit. A quick look at the source data indicates the more likely scenario is the latter.

What about the other organizations? How do they compare? We now need to solve equivalent linear programs for each one, so we can get the full picture. We do this by wrapping up everything we did above into a single function and then using the map() function we learned in Chap. 5 to iterate over each organization.

6.4.5   Create a Linear Program to Iterate for DEA

We'll create a function with two main arguments. The first argument is for the ith organization in the dataset, whose productivity portrayal the linear program will be optimizing. The second argument is the full input-output dataset itself. There's an additional argument in the fiction, set to zero by default, called min_weight. There may be circumstances where we don't want to let the algorithm assign a zero weight to any data element—insinuating that all elements should be emphasized at least to some extent. We can see from the results above that the algorithm gave x_staff a value weight of zero for Org A's optimal portrayal. For our purposes we won't adjust the min_weight argument because the DEA value weights are not as intuitive as the TI value weights, and setting min_weight too high at first go can break the algorithm. Further, the zero weights can be informative in themselves. You can play with the min_weight argument later once we cover more important matters of interpreting the base results in the reflection section below and in the next chapter.

```
# specify function arguments
optimize_dmu_output <- function(i, df, min_weight = 0) { # i refers
to the ith dmu

    # get data characteristics
    N <- nrow(dea_data_01)
    dmu_names <- dea_data_01$dmu
    inputs <- select(dea_data_01, starts_with("x"))
    outputs <- select(dea_data_01, starts_with("y"))
```

```r
  x_dim <- ncol(inputs)
  y_dim <- ncol(outputs)
  var_names <- c(colnames(outputs), colnames(inputs))

  # set all lp parameters
  objective <- c(as.numeric(outputs[i,]), rep(0, x_dim))
  aux <- cbind(as.matrix(outputs), as.matrix(-1*inputs))
  variable_constraints <- rbind(aux[-i,],
                      diag(x_dim + y_dim),
                      c(rep(0, y_dim), as.matrix(inputs[i,])))
  rhs_constraints <- c(rep(0, N-1), rep(min_weight, x_dim + y_
dim), 1)
  const_dir <- c(rep("<=", N-1), rep(">=", x_dim + y_dim), "=")

  # solve lp for ith dmu
  dmu_optimum <- lp(direction = "max",
                    objective.in = objective,
                    const.mat = variable_constraints,
                    const.dir = const_dir,
                    const.rhs = rhs_constraints)
  # get results
  results <- tibble("dmu" = rep(dmu_names[i], 1 + x_dim + y_dim),
                "variable" = c("Optimal Efficiency", paste("Weight",
var_names)),
                             "value" = c(dmu_optimum$objval,
dmu_optimum$solution)
                             )

  return(results)
}


save(optimize_dmu_output, file = "data/optimize_dmu_output.rda")
```

Now let's run this function for each organization. Recall how the map() family of functions works. It needs to iterate a function over a list of elements. Those elements must be inputs to the function that is being iterated. The first argument of map() is the list of elements to iterate over. The second argument will be the function we are running multiple times, and the last argument will be the full data set. We'll be using map_dfr() because we want a single data frame output that constitutes each iteration of the function bound together by rows.

```r
# sequence of DMU indices
dmu_indicies <- 1:dim(dea_data_01)[1]

# map it!
```

```
dea_base_results <- map_dfr(dmu_indicies, optimize_dmu_output,
                        df = dea_data_01)

# save for use in next chapter
save(dea_base_results, file = "data/dea_base_results.rda")

# check results
glimpse(dea_base_results)
## Rows: 36
## Columns: 3
## $ dmu <chr> "Org_A", "Org_A", "Org_A", "Org_A", "Org_A",
"Org_A", "Org...
## $ variable <chr> "Optimal Efficiency", "Weight y_participants",
"Weight y_c...
## $ value <dbl> 0.9016999093, 0.0005813690, 0.0075079329,
0.0007461469, 0....
```

We can see by the number of columns that the DEA algorithm has produced results for each of the six DMUs. Let's now put the results into wide format so we can interpret the results (Table 6.3).

```
# tidy up for readability
printable_results <- dea_base_results %>%
  mutate(value = round(value, 3)) %>%
  pivot_wider(names_from = "dmu", values_from = "value")

printable_results
```

The results are striking. First, when you build a DEA algorithm from the ground up like we have here, these are the first results you get. But if you use most off-the-shelf DEA packages, these are the first results you will miss. Most R packages do not report these value weights in their default results. Instead they give you something called lambda values, or shadow prices. We're not going to dig into lambda values in this chapter. Shadow prices are well documented in economics and linear programming literature, and they can be extremely helpful when analyzing

Table 6.3 DEA results for all organizations

Variable	Org_A	Org_B	Org_C	Org_D	Org_E	Org_F
Optimal efficiency	0.902	1.106	1.188	1.091	1.033	1.046
Weight y_participants	0.001	0.000	0.004	0.000	0.000	0.005
Weight y_certs	0.008	0.008	0.000	0.008	0.007	0.001
Weight x_sqrft	0.001	0.000	0.000	0.003	0.001	0.000
Weight x_staff	0.000	0.097	0.000	0.000	0.153	0.200
Weight x_workshops	0.028	0.021	0.050	0.000	0.000	0.000

for-profit firms. But with our NFPs, while they may have some explanatory value, they do not reveal nearly as much as the variable weights and the alternative optimal efficiency scores given in the table above.

The second reason why these results are profound is because of the actual Optimal Efficiency values in the table. Notice that all but one organization has a score greater than 1.0. This means that there is way to portray four out of five of these organizations as the most productive in the field. Recall that each time the DEA algorithm is run for an individual organization, it constrains the value of every other organization to a value of less than or equal to 1. Under these circumstances, the objective organization may be assigned a value greater than 1. The question then becomes, how should we think about these organizations' productivity with respect to one another in a clear, consistent, and helpful way? We will address this question with an analysis of the results in the next chapter.

6.5   Reflections on DEA in Practice

DEA is powerful, but its use and appreciation are often limited to niche circles of highly trained professionals. This situation represents both its strength and its tragic flaw. When you bring something so high powered into a context, such as education or the not-for-profit sector—which rarely relies on such mathematical sophistication for decision-making—the results of which can be admired, but DEA outputs in themselves don't often strike a chord for those with the agency to bring about change.

We thus take a sharp turn off the path that a common DEA application would take for using and interpreting results. At this point, based on coded-in decision rules largely grounded in economic theory and standard practices in operations research, most DEA algorithms would automatically choose which value weighting scheme to use for plotting a production frontier and determining the relative efficiencies between organizations. It is thus entirely possible that the DEA user may, not only be unaware of the criteria upon which the automated decision was made, but also be unaware that any decision point was reached. The implication is that most DEA algorithms settle on a specific set of results, which are then presented as definitive, and most of the information produced from having run the linear program is discarded without any consideration by the analyst.

This is precisely the reason why DEA algorithms will sometimes settle on a set of results when being run on education and not-for-profit data that simply do not make sense to someone who is highly familiar with that context and who has a different set of values than the ones programmed into the DEA's decision rules. As stated in the beginning of the chapter, we are taking a first principles approach to DEA because most applications of DEA have been developed outside the education sector. And please appreciate that the education context is one that often rejects the idea that increased efficiency is always a "good thing". This is why we take the time to deconstruct all intermediate results of the DEA algorithm and use judgement to determine which are most useful.

Values Analysis with DEA

7

7.1 Objective

In this chapter we raise our level of competency with the use of DEA. We harness the power of a foundational application of the technique[1] to generate actionable insight for NFP operational decision-making. By the end of the chapter, we will have accomplished three things. First, we'll learn how to take all of DEA's valuable information into account. Second, we'll produce specific types of insights that are needed to inform decision-making around NFP's nonfinancial objectives. Lastly, we'll learn how to visualize DEA's results for clear interpretation and presentation.

7.2 What Is Values Analysis with DEA?

With DEA values analysis, we step through the individual value weighting schemes produced by the DEA algorithm and test them against the practical and contextual circumstances of the organizations on the ground. The analysis can underscore what is working among a field of organization, what isn't, and whether performance aligns with key stakeholder interests. DEA value weight schemes serve as a characterization of an organization's de facto operating model (production function). Solutions to the DEA model can thus be tested and compared across the field of organizations to generate insight on relative performance and illuminate how performance gaps might be closed on an organization-by-organization basis.

Just like with TI values analysis, we use the numerical values of the input and output weights to reflect true human values. Procedurally, we test the sensitivity of end results to changes in the value weighting scheme of the inputs and outputs,

---

[1] Charnes, A., Cooper, W. W., & Rhodes, E. (1978). Measuring the efficiency of decision making units. European journal of operational research, 2(6), 429-444.

© The Author(s), under exclusive license to Springer Nature
Switzerland AG 2021
K. Moore, *Measuring Productivity in Education and Not-for-Profits*,
Management for Professionals, https://doi.org/10.1007/978-3-030-72965-3_7

which compose the productivity index. We do so in an informed way that considers (A) the operational circumstances and context of the NFP, (B) the social mandate or strategic objectives of the NFP, and (C) key stakeholder interests.

The prime difference between DEA values analysis and TI threshold analysis is that in this chapter we are not presenting plausible performance portrayals based on judgment. We will work through a similar process, but we'll be using empirically generated value weighting schemes produced by the DEA algorithm.

The approach has both drawbacks and advantages. First we highlight the drawback. As we learned in the previous chapter, the mathematical formulation of DEA serves to optimize the productivity ratio of each organization to highlight its most favorable performance portrayal. The optimal portrayal, however, may not practically make sense regarding the delivery of the service on the ground. Hence, while DEA-generated weighting schemes are both empirical and objective, unfortunately the numerical values of individual input and output value weights often stray from what is plausible or reasonable in a practical sense. For example, in order to give a multi-input, multi-output organization its most favorable performance portrayal, some of the variables must be given a value weight of zero. If the variable given a zero value weight is in fact essential to the production process, then any results taken at face value will not be helpful when it comes to decision-making.

On the other hand, the advantage of DEA values analysis is that the optimal productivity descriptions produced by the algorithm can be very telling regarding individual organizations' strengths and weaknesses. Because each organization is given its most complementary performance portrayal, areas where the organizations are excelling are highlighted. And conversely, areas where each organization is performing relatively poorly compared to their peers are also highlighted.

7.3   A Detailed Look at Values Analysis with DEA

7.3.1   The Procedure

Unlike in Chap. 6, we don't need to cover any more mathematical theory. Let's get straight into what we'll accomplish in this chapter.

1. We're going to interpret the full set of results generated by the DEA algorithm. Practically, this means making sense of the unique value weighting schemes given to each organization, which serve to optimize each of their productivity portrayals. Similar to what we have done in previous chapters, we'll need to slightly transform the data and normalize the figures, so they are more intuitive for interpretation and communication.

2. Once we have visualized all of the DEA's solutions to the optimization problem, we'll relate those solutions back to the real-world operational contexts of these organizations. We'll see what each set of solutions means practically for their respective organizations, and we'll consider what human values are likely at play for quantitative value-weighting scheme.

3. With a clearer understanding of what each value weighting scheme represents, we can then test each organization against each set of values. Recall that the

DEA algorithm only shows Organization A's productivity score for its own optimal value weighting scheme. But how might Org A's performance look if it were judged under the criteria that gave Organization B its optimal productivity score? We'll calculate all the different combinations in this step.

4. Next, we'll visualize performance among the field for each value weighting scheme, and with a clearer understanding of what values are being emphasized in each scheme, we can make a clearer judgment about what relative performance between the organizations truly looks like.

5. Finally, we'll entertain the possibility that none of the computer-generated value weighting schemes appropriately encapsulates our values for how performance should be judged. If such is the case, we'll work through a few different options that might still produce some acceptable and compelling results.

7.4   Tutorial: DEA Values Analysis for Decision-Making

7.4.1   Setup

```
# packages
library(knitr)
library(tidyverse)
library(scales)
library(ggrepel) # prevents overlapping text in charts
library(glue)

# data
load("data/dea_data_01.rda")
load("data/dea_base_results.rda")
load("data/optimize_dmu_output.rda")
```

7.4.2   Introduction

Let's first expand upon the scenario presented in the previous chapter, where we are helping a government agency conduct a process evaluation of six organizations contracted to delivery training workshops in six different locations. The reason why the process evaluation has been commissioned is because the training service providers' quarterly reports have indicated that two distance approaches to implementing the trainings have emerged. Organizations A, B, and C (Group 1) are employing a strategy that involves running as few workshops as possible but recruiting as many participants as possible for each one. Organizations D, E, and F (Group 2) are perusing a strategy where they run more workshops with fewer participants in order to provide better quality assurance and ensure a more positive participant experience.

Our objective is to use these insights from the quarterly reports to determine which organizations are performing best against their own objectives and also to

Table 7.1 Review of DEA results

variable	Org_A	Org_B	Org_C	Org_D	Org_E	Org_F
Optimal efficiency	0.902	1.106	1.188	1.091	1.033	1.046
Weight x_sqrft	0.001	0.000	0.000	0.003	0.001	0.000
Weight x_staff	0.000	0.097	0.000	0.000	0.153	0.200
Weight x_workshops	0.028	0.021	0.050	0.000	0.000	0.000
Weight y_certs	0.008	0.008	0.000	0.008	0.007	0.001
Weight y_participants	0.001	0.000	0.004	0.000	0.000	0.005

determine which approach aligns best with the government agency's interests in approving this new training program in the first place.

First, let's recall the DEA results from the previous chapter. Pay special attention to the top-performing institutions in each group (Table 7.1).

The table above suggests that for Group 1, Organization C is the most productive and for Group 2, Organization D is the most productive. Within each group, these are the organizations whose optimal performance portrayals set them farther apart from their peers. So as we proceed with the analysis, we want to pay special attention to Organizations C and D.

7.4.3   Prepare the Data for Values Analysis

First we need to get all our data in the same place. Let's combine our DEA results with the initial inputs and outputs data set. We do this because each time we ran the DEA algorithm, we produced productivity scores only for the organizations whose productivity portrayal was being optimized. We didn't, however, get the scores for the organisations who weren't the objective of the optimisation algorithm, i.e. those serving as the constrains. We can easily find their productivity scores by using the weights provided in the original performance data.

First, let's isolate only the value weights from the DEA output.

```
# filter out the optimal efficiency scores, and
# rename some variables for more clarity
results_weights <- dea_base_results %>%
  filter(variable != "Optimal Efficiency") %>%
  mutate(variable = str_remove(variable, "Weight "),
         dmu = as.character(dmu)) %>%
  rename(opt_weight = value) %>%
  arrange(dmu, variable)

# view result
glimpse(results_weights)
## Rows: 30
## Columns: 3
```

```
## $ dmu         <chr> "Org_A", "Org_A", "Org_A", "Org_A", "Org_A",
"Org_B", "O...
## $ variable    <chr> "x_sqrft", "x_staff", "x_workshops", "y_
certs", "y_parti...
## $ opt_weight <dbl> 0.0007461469, 0.0000000000, 0.0276155938,
0.0075079329, ...
```

We've now got a simpler data frame indicating the optimal weights for each variable for each DMU. Next, pivot the initial dataset to long format, and combine with the DEA output.

```
# reshape initial data set
dea_data_long <- dea_data_01 %>%
  pivot_longer(x_sqrft:y_certs, names_to = "variable", values_to =
"value") %>%
  arrange(dmu, variable)

# join together and organise for clarity
dea_all_data <- dea_data_long %>%
  left_join(results_weights) %>%
  separate(variable, c("type", "variable"), sep = "_")

glimpse(dea_all_data)
## Rows: 30
## Columns: 5
## $ dmu         <chr> "Org_A", "Org_A", "Org_A", "Org_A", "Org_A",
"Org_B", "O...
## $ type        <chr> "x", "x", "x", "y", "y", "x", "x", "x", "y",
"y", "x", "...
## $ variable    <chr> "sqrft", "staff", "workshops", "certs", "par-
ticipants", ...
## $ value       <dbl> 600, 6, 20, 105, 195, 775, 6, 20, 132, 240,
800, 7, 20, ...
## $ opt_weight <dbl> 0.0007461469, 0.0000000000, 0.0276155938,
0.0075079329, ...
```

We've now got all of our data in one place—the original input-output data and the DEA value weights that optimize the productivity portrayal for each organization. Our next step is to begin understanding the significance of each optimal value weighting scheme.

7.4.4   Production Function Analysis

Better understanding about the DEA weights will give more clarity to how each organization achieved its most favorable productivity score. The weights serve to emphasize certain variables. In other words, the weighting scheme for the objective organization tells us the areas where that organization is performing best, relative to its peers. Conversely, the variables that are not emphasized are likely where the other organizations are performing better. Outputs are intuitive. For example, if the objective organization has relatively greater outputs in one area, DEA will want to give a greater value weight to that variable. The greater value weight will ensure the objective organization's performance in that area is emphasized, leading to a more desirable productivity score. For inputs, the method is the same, but the logic is counter intuitive. Inputs appear in the denominator of the productivity ratio, so fewer inputs are more desirable. If the objective organization has relatively fewer inputs in one area, DEA will also want to give that variable a greater value weight— but not because it wants to increase the value of that variable for the objective organization. Rather, since the other organizations have relative greater values for that variable, the DEA weights will serve to make their productivity scores proportionally less desirable.

If we are going to better understand the significance of the different weighting schemes, we need to put them in a form that is easier to understand. We want to easily communicate the emphasis that the DEA algorithm places on each input-output variable for all organizations' optimal productivity portrayals. We'll need to transform the data slightly to best communicate results by normalizing them to a scale that's easier to understand. We'll normalize the value weights so that each organization's output emphasis scores will sum to 100%, and likewise each organization's input emphasis score will 100% (Table 7.2).

```
# normalise and rescale the weights by type
weight_emphasis <- dea_all_data %>%
group_by(dmu, type) %>%
 mutate(emphasis = opt_weight /sum(opt_weight)) %>%
ungroup() %>%
select(-opt_weight, -value)
```

Table 7.2 Organization A's value weighting scheme

dmu	Type	Variable	Emphasis
Org_A	x	sqrft	0.5068140
Org_A	x	staff	0.0000000
Org_A	x	workshops	0.4931860
Org_A	y	certs	0.9281311
Org_A	y	participants	0.0718689

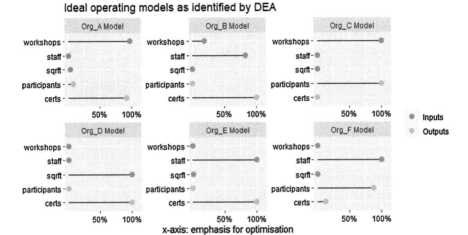

Fig. 7.1 Visual comparison of the different DEA weighting schemes

```
# view results for Org A
filter(weight_emphasis, dmu == "Org_A")
```

Results for Organization A tell us that, for inputs, the DEA algorithm has given equal emphasis to the size of the training facility and the number of workshops delivered. For outputs, however, far greater emphasis is placed on the number of certifications awarded than the number of participants the workshops attract. Let's now visualize the results for the whole field of organizations, so we can better interpret what's happening with this data (Fig. 7.1).

```
plot_emphasis <- function(df) {


  p <- df %>%
    ggplot(aes(x = variable, y = emphasis)) +
    geom_segment(aes(x=variable, xend=variable, y=0, yend=emphasis),
color = "black", size=0.5) +
    geom_point(aes(color=type), size=3, shape = "circle") +
    coord_flip() +
    facet_wrap(~ dmu, scales = "free",
                    labeller = labeller(dmu = function(x){paste(x,
"Model", sep = " ")})) +
    scale_color_discrete(labels = c("Inputs", "Outputs")) +
    scale_y_continuous(labels = function(x) {scales::percent(x,
accuracy = 1L)},
```

```
                         breaks = c(0.5, 1),
                         limits = c(0,1.1)) +
      theme(legend.title = element_blank(),
            axis.title.y=element_blank()) +
      labs(title = "Ideal operating models as identified by DEA",
           y = "x-axis: emphasis for optimisation") # remember coord_
flip above

return(p)
}


# run function
plot_emphasis(weight_emphasis)
```

The results show the areas of performance where each organization is excelling, or rather, which areas of performance when emphasized give the organization its most favorable productivity portrayal relative to the others. First direct
your focus toward Organizations C and D. Recall from the scenario introduction
section that we have some contextual information about each service provider's
strategy. We know that Organizations A, B, and C have been focusing more on
running fewer workshops but recruiting as many people to attend each one as possible, whereas Organizations D, E, and F have been focusing more on training
quality. Further, our DEA results show that Organization C was the most productive among the first group and D was most productive among the second group.
The chart above further captures these two approaches. The figure portraying Org
C's optimal operating model suggests that emphasizing the ratio
"participants"/"workshops" gives results in its most favorable portrayal. That is,
relative to the other organizations, it is doing a better job at minimizing the number of workshops and maximizing the number of participants. Notice how
Organization D's operating model is different. Org D has excelled in creating a
more intimate space with a smaller training facility and converting workshop par-
ticipation into workshop completions with a certificate.

At this point you might ask yourself, why don't Organization A and B's operating model more resemble Org C's, since they are pursuing similar strategies. And
why don't Org E and F's model more resemble Org D's? A simple answer could be
that they have not been as effective at implementing their strategies. The strategy
information is a key piece of contextual information we can use in our analysis to
interpret results, but the DEA algorithm gives a context-free assessment of how

Table 7.3 Optimal productivity scores for each organisation

Variable	Org_A	Org_B	Org_C	Org_D	Org_E	Org_F
Optimal efficiency	0.902	1.106	1.188	1.091	1.033	1.046

each organization maximizes its productivity portrayal. There will always be more and less productive organizations and more and less effective organizations at implementing their strategies. Organizations C and D have shown to epitomize their respective strategies, and it shows up in the data.

7.4.5   Values Analysis for Examining the Production Frontier

We have interpreted some of the most significant results of our DEA analysis. A few questions remain, however. How do we best compare these organizations on a level playing field? How can we best characterize performance gaps between them? And which of the operational models above best reflects the funder's intentions in commissioning these training programs in the first place? These questions are at the heart of our values analysis.

Let's frame the first question regarding comparisons on level playing field. Recall the optimal efficiency scores determined by the DEA algorithm (Table 7.3).

These scores are determined, essentially, by allowing each organization to play by its own rules. The weighting schemes that produce these results are different for each organization and thus represent a different set of values being emphasized for each organization. This means we have to be cautious in taking the results at face value. We shouldn't assume, for example, that all else held equal, Organization C is the most productive with a score of 1.188. That may not be the case in practice or in principle.

Above we examined multiple operational models that emphasize different aspects of performance for each organization. What if we selected one or two of these models that best represent what the organizations are set up to achieve and test how each of the organization performs under that single value weighting scheme. For example, Organization C significantly outperforms the other organizations when emphasizing its own strengths, but what if we look at Org C's performance through the lens that gave Organization D its most favorable productivity portrayal? This is how we can begin addressing the second and third questions raised above regarding real performance gaps and assessing performance based on the funder's intentions, rather than by how each organization would prefer to be judged.

What we need to do is build a more comprehensive data set that computes each organization's productivity based on each of the different operating models covered above. To do this we'll design a function that isolates one of organization's optimal value weighting schemes and then computes productivity scores for each organization against that scheme. We'll then iterate that function using map() for each of the different value weighting schemes.

```
# define function parameters
alt_scores <- function(org, df) {
```

```
# number of DMUs
N <- length(unique(df$dmu))

# isolate one value weighting scheme
org_scheme <- df %>%
  filter(dmu == org) %>%
  .$opt_weight

# apply that scheme to each of the other DMUs
org_scheme_scores <- df %>%
  mutate(opt_weight = rep(org_scheme, N),
         scores = value*opt_weight) %>%

  # find the numerator (output) and denominator (input) indexes
  # for the productivity ratio using the value weights
  group_by(dmu, type) %>%
  summarise(index = sum(scores)) %>%
  ungroup() %>%

  # divide outputs by inputs to find productivity
  pivot_wider(names_from = type, values_from = index) %>%
  mutate(prod_index = y/x,
         scheme = paste(org, "scheme", sep = "_")) %>%
  select(scheme, everything())

  return(org_scheme_scores)
}


# list all DMUs to iterate the function
dmu_names <- unique(dea_all_data$dmu)

# map function for each dmu
all_scores <- map_dfr(dmu_names, alt_scores, dea_all_data)

# check result
glimpse(all_scores)
## Rows: 36
## Columns: 5
## $ scheme      <chr> "Org_A_scheme", "Org_A_scheme", "Org_A_
scheme", "Org_A_s...
## $ dmu         <chr> "Org_A", "Org_B", "Org_C", "Org_D", "Org_E",
"Org_F", "O...
## $ x           <dbl> 1.000000, 1.130576, 1.149229, 1.190390,
1.365775, 1.4030...
```

Table 7.4 Weighting schemes for each organization

dmu	Scheme_A	Scheme_B	Scheme_C	Scheme_D	Scheme_E	Scheme_F
Org_A	0.90	0.80	0.68	0.36	0.51	0.77
Org_B	1.00	1.00	0.84	0.35	0.59	0.95
Org_C	1.00	0.90	1.00	0.34	0.53	0.96
Org_D	1.00	0.90	0.35	1.00	0.97	0.86
Org_E	0.94	0.89	0.34	0.92	1.00	0.96
Org_F	0.91	0.89	0.36	0.80	0.97	1.00

```
## $ y          <dbl> 0.9016999, 1.1305757, 1.1492294, 1.1903898,
1.2770966, 1...
## $ prod_index <dbl> 0.9016999, 1.0000000, 1.0000000, 1.0000000,
0.9350709, 0...
```

Just as with TI, we've isolated the input and output indexes, as well as a final productivity index for each organization under each value weighting scheme produced by the initial DEA. The individual input and output indexes will allow us to later plot these organizations on a 2-D plane with a frontier, so that we can compare them to one another. First, though, let's reshape the data, so we can get our heads around the results. Just as in previous chapters, we'll use some feature scaling so that visualization and interpretation of results are more intuitive. We need to do this because, as you can see in the data, even though the DEA algorithm sets a constraint on all other organizations that productivity cannot be greater than 1, sometimes the productivity score for the organization being optimized ends up marginally greater than 1. Normalizing from here just makes results more intuitive, and the variation in the data is still preserved (Table 7.4).

```
# Calculate technical efficiency scores, so
# all orgs fall between 0 and 1, and
# pivot to wide format
efficiency_schemes_matrix <- all_scores %>%
  select(-c(x,y)) %>%
  group_by(scheme) %>%
  mutate(TE = prod_index/max(prod_index)) %>%
  ungroup() %>%

  # rename for brevity
  mutate(TE = round(TE, 2),
         scheme = str_replace_all(scheme, c("Org_A_scheme" =
"Scheme_A",
                                            "Org_B_scheme" =
"Scheme_B",
                                            "Org_C_scheme" =
"Scheme_C",
                                            "Org_D_scheme" =
"Scheme_D",
                                            "Org_E_scheme" =
"Scheme_E",
                                            "Org_F_scheme" =
"Scheme_F"))) %>%
  select(-prod_index) %>%
  pivot_wider(names_from = scheme, values_from = TE)

efficiency_schemes_matrix
```

The efficiency matrix lists the technical efficiency scores for each organization under each value weighting scheme. If we look at the results from Schemes C and D, we can see the stark difference in how productivity is portrayed. Under Org D's scheme, Org C performs at the bottom and vice versa.

We'll now direct our attention back to the all_scores data set. Using the x and y indexes, we'll create a visualization for these organizations that defines a production frontier to get a clear visual picture of the implications of the alternative value weighting schemes on performance portrayals.

There are a few insights and details worth noting before we draw our plots. First, we'll chart the organizations on an x-y plane, and the frontier will emanate from the origin. Since we are defining the frontier based on the portrayal of the highest performing organization under each scheme, we will draw a line from the origin through the point of the highest performing organization to visualize the frontier. Any position below the frontier line signifies a lower level of outputs (y-index), and any position to the right of the frontier line signifies a higher level of inputs (x-index). Both circumstances mean a lower level of productivity than the frontier.

7.4.6   Visualize the Frontiers

First, create a plotting function to visualize results for one weighting scheme, and then use the map() function to create a plot for each scheme.

```
# define function parameters to make plot for
# a single value weighting schemes
plot_frontiers <- function(sch, df) {
```

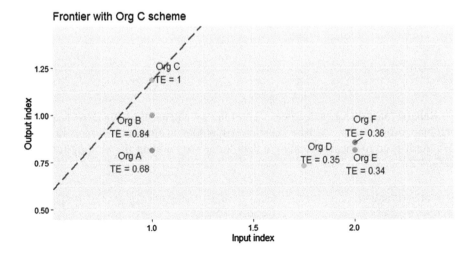

Fig. 7.2 Productivity frontier for Organization C's weighting scheme

```
# normalise results, as we did in the efficiency matrix, so
# all productivity scores will fall between 0 and 1
data <- df %>%
  filter(scheme == sch) %>%
  mutate(TE = prod_index/max(prod_index))

# find slope of frontier line
frontier_slope <- data %>%
  filter(TE == max(TE)) %>%
  filter(y == max(y)) %>%
  mutate(slope = y/x) %>%
  .$slope

# plot organizations against the frontier
plot <- data %>%
  ggplot(aes(x = x, y = y)) +
  geom_point(aes(color = dmu), size = 3) +
  geom_abline(intercept = 0, slope = frontier_slope, size = 1,
alpha = 0.7, linetype = 5)  +
    geom_text_repel(aes(label = glue("{str_replace(dmu, '_', '
')}\nTE = {round(TE, 2)}"))) +
    # adjust the scale of the chart
    scale_y_continuous(limits = c(min(data$y)-0.2*(max(data$y)),
max(data$y)+0.2*(max(data$y)))) +
    scale_x_continuous(limits = c(min(data$x)-0.2*(max(data$x)),
max(data$x)+0.2*(max(data$x)))) +
    theme(legend.position = "none") +
    labs(title = glue("Frontier with {str_replace_all(sch, '_',
' ')}"),
         y = "Output index",
         x = "Input index")

  return(plot)
}
```

Run the function for each scheme. This time we will use the map() base function, not map_dfr() because the base function will automatically store all plots in a list. We don't need to specify creating a data frame for iterating the plotting function (Fig. 7.2).

```
# list schemes
schemes_list <- unique(all_scores$scheme)

# run function
frontier_plots <- map(schemes_list, plot_frontiers, all_scores)
```

```
# name the plots according to their scheme
names(frontier_plots) <- schemes_list
Now let's check out the results for Org C.
frontier_plots$Org_C_scheme
```

Let's quickly interpret results. Recall that from the operating model visualization above, Org C's value weighting scheme only considers "participants"/"workshops" for its productivity ratio. Each of Organization A, B, and C ran 20 workshops, so their level of input is identical. However, from those 20 workshops, Org C had the most participants. The other organizations did not excel with respect to only those two performance indicators, so they are shown to have both higher input and lower output.

Now check out the frontier plot for Org D (Fig. 7.3).

```
frontier_plots$Org_D_scheme
```

Organization D's ideal productivity portrayal considers only "Certifications"/"Square feet." While Org E and F awarded more certifications, they did so with much larger training facilities, i.e., less effective space utilization. The other organizations did not perform well against these indicators and so find themselves on the opposite end of the spectrum.

What if we wanted a more balanced perspective on performance? What might top performance look like under different conditions? There are two other organizations in the field that, when assessed on their own ideal criteria, they also outperformed everyone else. As the evaluator, we should use the opportunity to refer to the funder's priorities. Let's say the funder is predominantly concerned about the number of certificates awarded. That is, they want as many qualified individuals

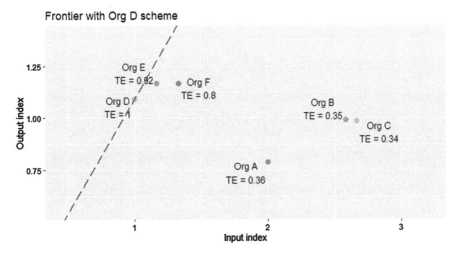

Fig. 7.3 Productivity frontier for Organization D's weighting scheme

completing the trainings as possible. Further, they aren't interested in judging the organizations based on their fixed costs. They don't see the type or size of the facility as a significant performance factor. They recognize that staff salaries and workshop delivery represent the most significant direct expenditures. Under these circumstances, a quick look back at the operational models tells us that Org B's scheme most closely represents the funder's values. And if we plot all organizations with respect to Org B's weighting scheme, we get yet another, very different performance portrayal (Fig. 7.4).

Under Org B's scheme, no single organization or group of organizations is too far ahead or too far behind. Instead, what we see is that when compared to Org B's achievements, most other organizations are running at about 90% efficiency. Under this set of performance criteria, one concussion the funder might make is that scaling up operations is a difficult task. Other organizations have spent more on inputs and have even awarded more certificates, but they have done so less efficiently and effectively. If we are satisfied with this portrayal, then a clear next step would be to make a case study out of Org B and see how they have achieved this level of performance on the ground.

7.4.7   What Happens When No DEA Scheme Represents Our Values?

It's possible that the DEA algorithm produces results, none of which reflect what we are trying to achieve. That is, when you examine how the DEA value weights emphasize individual variables in the productivity calculation, none of those value weighting schemes seem to align with our strategic objectives. The implication is that if you proceed with the analysis and end up choosing one or two of these

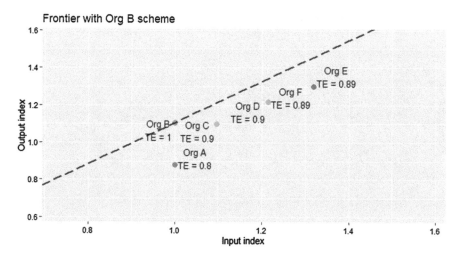

Fig. 7.4 Productivity frontier for Organization C's weighting scheme

schemes to plot a frontier, that frontier will not be accurate, and the relative organization performance portrayals won't be appropriate.

There are a few directions we could go from here. First, we may just decide that defining a frontier and using it as a reference to judge relative performance is just not the way to go. This doesn't mean, however, that the results that have gotten us to this point are not valuable. Not only do the DEA value weights highlight where each individual organization is excelling; the variables the algorithm did not emphasize for each organization represent areas where they are not performing as well as their peers. The weighting schemes alone serve as strong anchor points for further investigation. They immediately raise questions about why and how about the data.

Second, we could define a customized value weighting scheme based totally on judgment or define a range of acceptable weights, much as we did in Chap. 5. We'll look at an example of how this can be done for absolute productivity measures in Chap. 8. It will build from the TI techniques we've already covered.

Third, we can tweak the DEA algorithm. For example, let's say we put off that the DEA algorithm was so selective in the variables it chose to emphasize for each organization, and we really want to make sure that we define optimal performance such that all performance variables are taken into account at least to some extent. What we can do here is set some minimum acceptable value for each variable. We must be careful about how we do this, however. The raw DEA value weights are not very intuitive, but we can look back at our first DEA results to find a place to start (Table 7.5).

We can see that under no restrictions, the DEA algorithm gave several different variables a value weight of 0.001. In our first attempt, we wouldn't want to set our minimum value above this weight because it could make the linear program unsolvable. So, let's rerun the algorithm with a minimum weight of 0.001, so each variable is accounted for to at least some meaningful extent (Fig. 7.5).

```
# list dmus
dmu_indicies <- 1:nrow(dea_data_01)

# map function for all dmus
dea_new_results <- map_dfr(dmu_indicies, optimize_dmu_output,
                           df = dea_data_01,
                           min_weight = 0.001)
```

Table 7.5 Review of DEA full results

Variable	Org_A	Org_B	Org_C	Org_D	Org_E	Org_F
Optimal efficiency	0.902	1.106	1.188	1.091	1.033	1.046
Weight x_sqrft	0.001	0.000	0.000	0.003	0.001	0.000
Weight x_staff	0.000	0.097	0.000	0.000	0.153	0.200
Weight x_workshops	0.028	0.021	0.050	0.000	0.000	0.000
Weight y_certs	0.008	0.008	0.000	0.008	0.007	0.001
Weight y_participants	0.001	0.000	0.004	0.000	0.000	0.005

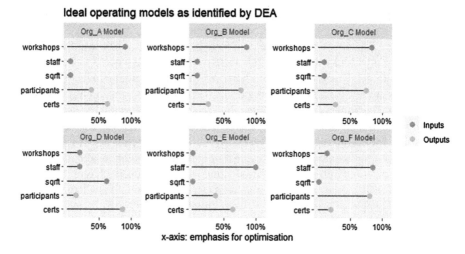

Fig. 7.5 Alternative weighting schemes for different DEA parameters

```
# organise and combine with org data
results_weights_new <- dea_new_results %>%
  filter(variable != "Optimal Efficiency") %>%
  mutate(variable = str_remove(variable, "Weight "),
         dmu = as.character(dmu)) %>%
  rename(opt_weight = value) %>%
  arrange(dmu, variable)

dea_all_data_new <- dea_data_long %>%
  left_join(results_weights_new) %>%
  separate(variable, c("type", "variable"), sep = "_")

# see the new weight emphasis
weight_emphasis_new <- dea_all_data_new %>%
  mutate(emphasis = ifelse(type == "x" & opt_weight != 0, 1 - opt_
weight, opt_weight)) %>%
  group_by(dmu, type) %>%
  mutate(emphasis = emphasis/sum(emphasis)) %>%
  ungroup() %>%
  select(-opt_weight, -value)

# Examine new op models
plot_emphasis(weight_emphasis_new)
```

When all variables must be considered to optimize each organization's performance portrayal, we get much less variability in what the ideal operational model looks like. With only some minor variations, the algorithm has determined that each input variable is of equal importance. The output weights, however, still get very different treatments depending on what we want performance to look like. Let's say we want to emphasize the awarding of certificates more than simply attracting participants. This narrows our choices down to Org A, Org D, and Org E's models. We also would like to give the organizations as much flexibility with staffing as possible, so we like the fact that Org E's model emphasizes the staff input slightly less than the others. In this case, let's check out what productivity looks like under this scheme (Fig. 7.6).

```
# get all scores for all weighting schemes
all_scores_new        <-       map_dfr(dmu_names,       alt_scores,
dea_all_data_new)

# plot all frontiers
frontier_plots_new    <-       map(schemes_list,       plot_frontiers,
all_scores_new)
names(frontier_plots_new) <- schemes_list

# check out Org D's new frontier plot
frontier_plots_new$Org_E_scheme
```

We still have two clear groups, aligning with what we'd expect from the information provided about these organizations in the introduction section, where their quarterly reports indicated that they took different strategic directions. The performance portrayals are more forgiving, however, than when we chose Org D's scheme in the first DEA iteration above when only two performance indicators were considered. To experiment further with the data, go to ikenresearch.com/interact.

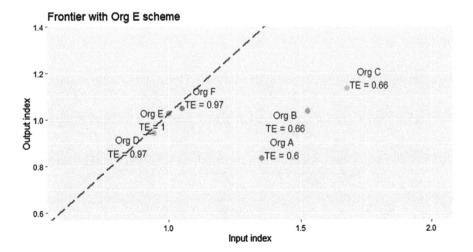

Fig. 7.6 New frontier for Org E

7.5   Reflections on DEA Values Analysis in Practice

The applications of DEA we have worked through in this chapter and in the previous one were performed on relatively simple data, but the applications can be scaled as complexity increases. For example, NFPs often don't employ a permanent set of full-time staff. They may not have a facility or even an established office that represents a significant productive asset with respect to the service being delivered. Yet demonstrating success through effective and efficient operations is still important. In such cases alternative inputs and outputs can be explored, such as numbers of different types of staff, numbers of site visits, numbers of referrals to other organizations, or some other representation of having delivered upon a task.

NFP leaders and managers often shy away from reporting hard numbers against aspirational goals, and they may even be uncomfortable with setting targets against shorter-term objectives. The apprehension is understandable. If your NFP has been set up, for example, to combat a certain health condition across a large remote region of a country, then what are the chances that your isolated intervention can produce demonstrable shifts in population health indicators over the course of a 2-year grant? Not very likely.

The irony, however, is that not having any operational data to scrutinize at all usually represents a much worse situation for the NGO than reporting data that is less-than-ideal. If no data are presented, then critical arguments for why the funder should consider allocating funding elsewhere can only be combatted by other qualitative arguments or testimonials about the success of the service. The author of this text has served as a program evaluator for several organizations on behalf of a funder, and no matter the direction of observed trends in data, empirical evidence has always been the service provider's friend. When NFP leaders and managers are reluctant to share or even collect reliable data on operations, the story almost never ends well for them. Real productivity data that can be monitored, explored, and leveraged more often creates opportunities for NFPs to improve than condemns their mistakes.

Threshold Analysis for Absolute Productivity

8

8.1 Objective

This chapter extends applications of the Törnqvist Index (TI) for measuring levels of productivity between individual organizations across a field of peers. By the end of this chapter, you will have practiced an alternative method to DEA for analyzing absolute productivity. If DEA methods seem too deterministic or just too technically complicated, then this chapter will provide a solution for measuring absolute productivity, which allows the researcher to define value weight ranges in the same manner as we did for productivity change in Chap. 5.

8.2 What Is Absolute Productivity Threshold Analysis?

Threshold analysis for absolute productivity applies productivity change scores across a period to estimates of absolute productivity in the base year. The prime difference you'll notice about the absolute productivity change threshold is that we don't default to an initial reference point of "1" for the base year. Instead the value weighting schemes also apply to initial conditions, and those initial conditions may be compared directly with other institutions. The figure below provides an example. Absolute productivity change scores may stoke more heated debate around the relative performance of different institutions, but they put the productivity change scores into more perspective (Fig. 8.1).

For example, if we measure productivity change with respect to a non-descript starting point, as in previous chapters, then true increases in productivity are difficult to judge. An institution that doubles its productivity over a period from a starting point of 0.25 increases its productivity to a level of 0.5. But an institution that doubles its productivity from a starting point of 2.0 increases its productivity to a level of 4.0. The magnitude of the difference in the real change in productivity in

© The Author(s), under exclusive license to Springer Nature   119
Switzerland AG 2021
K. Moore, *Measuring Productivity in Education and Not-for-Profits*,
Management for Professionals, https://doi.org/10.1007/978-3-030-72965-3_8

Range of plausible productivity trends

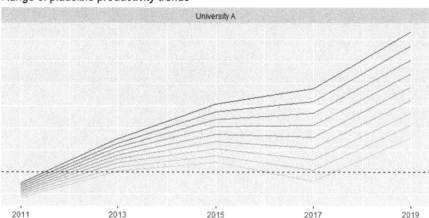

Fig. 8.1 Example visualisation of absolute productivity threshold analysis

this case is almost tenfold. The first institution jumped productivity by a real value of 0.25, while the second institution jumped productivity by a real value of 2.0. Yet using only traditional TI measures, each institution would have been shown to be equivalent, increasing its productivity by a factor of 2.

8.3   A Detailed Look at Absolute Productivity Threshold Analysis

8.3.1   Overview of the Procedure

A simple example of how the process works will best serve our purpose in explaining how absolute productivity threshold analysis can be used in conjunction with the productivity change threshold analysis we conducted in Chap. 5.

Consider one institution's cumulative productivity change scores over a 5-year period (Table 8.1).

Recall that when using TIs, the base year is always a value of 1. The cumulative change scores are always relative to the level of productivity at the beginning of the period, regardless of what that level of productivity is. The cumulative change scores from TI analysis simply allow for tracking the directionality of productivity change trends irrespective of levels of absolute productivity. They are primarily helpful for single institutions analysis, or understanding broad trends across a field of institutions. No comparison can be made between institutions regarding which one is more productive than another.

What if, however, we did have an estimate of the level of absolute productivity for our hypothetical institution above for the year 2011. We already know the extent to which productivity has changed over the period, so if we had a measure of absolute productivity at the beginning, we could easily find the level at the end.

Table 8.1 Example
university cumulative change

Institution	Year	cum_change
Uni X	2011	1.00
Uni X	2013	1.02
Uni X	2015	1.07
Uni X	2017	1.11
Uni X	2019	1.12

Table 8.2 Example university absolute productivity change

Institution	Year	cum_change	abs_prod_base	abs_prod_change
Uni X	2011	1.00	0.75	0.75
Uni X	2013	1.02	0.75	0.76
Uni X	2015	1.07	0.75	0.80
Uni X	2017	1.11	0.75	0.83
Uni X	2019	1.12	0.75	0.84

Let's say we already have a defined scale for interpreting levels of absolute productivity, and we've estimated that in 2011, Uni X's level of absolute productivity was 0.75. The cumulative productivity change scores simply represent factors of change since the base year. In the original TI threshold analysis, the reference level of productivity for any institution was normalized at a default level of "1." If we have evidence, however, of the actual level of productivity, we can just multiply the cumulative change scores against that base level to see how the absolute level productivity has changed across the period (Table 8.2).

We can immediately see that a 12% increase in productivity over the period brings a base level of productivity of 0.75 up to a new level of 0.84.

8.3.2   Estimating Absolute Productivity

Recall from Chaps. 6 and 7 that absolute productivity is often best communicated in terms of the concept technical efficiency (TE), which places all productivity scores from a field of organizations on a scale from 0 to 1, where the score of 1 is defined as the maximum. We performed an analogous operation as well as in Chap. 3, where we first introduced the idea of normalizing productivity scores to better communicate results.

In this chapter, we'll go through a process of feature scaling each element in the dataset with respect to the maximum value observed for that element. When we aggregate the data to get a productivity score, we'll have a scale of productivity that is oriented around a value of 1. The difference, however, between the new indicator and the TE indicator from the previous chapter will be that on the new scale, a value of 1 will not symbolize the maximum achievable level of productivity. Instead, since the inputs and outputs themselves will be scaled to a maximum value of 1, a productivity ratio of 1 will be equivalent to the maximum observed output divided by the maximum observed input. For the new productivity scale we'll create in this chapter, a score of 1 thus serves more as an arbitrary reference point. The important thing to remember is that productivity scores will all sit on a consistent scale. This means that regardless of being marginally above or below that reference point of 1, if an institution has a productivity score that is double another institution's score, then the first institution is assumed to be twice as productive as the second. That is, for a fixed level of input, the first is producing twice the output. Or conversely, for a fixed level of output, it is using half the input. And to reiterate, because all the individual elements will be feature scaled prior to calculating the productivity ratio (as is done with TI methods), the productivity scores will still be within an intuitive range, as to not distract from the insight that the analysis is intended to provide.

8.4   Tutorial: Teaching Research Nexus?

8.4.1   Setup

To run this script, we'll need to load the following packages and several objects from previous chapters.

```
# packages
library(knitr)
library(tidyverse)
library(glue)

# functions
load("data/change_factor.rda")
load("data/calc_change.rda")
load("data/plot_prod_change.rda")
load("data/prod_change_calc.rda")
load("data/get_ranks.rda")

# data
load("data/unis_adjusted.rda")
load("data/prod_range.rda")
```

8.4.2   Introduction

Higher education system leaders were pleased with the productivity change scores, which helped both to confirm and challenge some of their assumptions. Some of the results, however, were surprising and difficult to explain. Their contextual knowledge of Universities A and C have led them to think that the original TI indexes have overstated the gains they have made over the past 8 years. They would like a supplemental analysis that places all institutions on the same playing field, so that relative gains can be compared more directly.

8.4.3   Feature Scaling All Data Elements

Our first step is to place all indicators on a consistent scale. We do this so that when we aggregate inputs and outputs into a single ratio, data elements measured in individual persons, for example, aren't overshadowed by elements with figures in the millions of dollars. In Chaps. 4 and 5, the TIs we created for measuring change accomplished this step automatically. This time, however, since we want to estimate absolute productivity, we'll perform a normalization of each data element relative to the maximum value for that data element across institutions. As explained in Chap. 3, this feature scaling step places all data on a level playing field while preserving all the variation in the dataset.

We'll perform the feature scaling a single time for each type of variable measured in the same units. We do this to avoid overemphasizing or underemphasizing different data elements measured in the same unit relative to one another. The step we'll take to accomplish this is to first lump together any data elements measured in the same unit. We'll only have to do this for inputs. By doing this, lose the ability to determine the extent to which individual input variables influence the final productivity metric. For the purposes of this chapter, however, we do perform this operation for the sake of simplicity, recognizing that we could go back and treat each variable separately if further questions arise or further investigation is required.

We already have robust measures of productivity change for this data that we generated in Chap. 5, so we'll only perform the feature scaling to determine absolute productivity estimates for the base year, 2011. Once we have the absolute productivity estimates for that base year, we can use the productivity change scores we've already calculated in conjunction with the new estimates. We'll create a simple function to perform the scaling and normalization, in case it may come in handy for you in the future.

```
# scale data
data_for_scaling <- unis_adjusted %>%
   filter(year == 2011) %>%
   mutate(x = x_salaries + x_cap_ex + x_intermediate) %>%
   select(institution, year, x, y_adjusted_load, y_adjusted_pubs)

# create function to scale variables
scale_all_vars <- function(df) {

 data_scaled <- df %>%
   gather(key = "variable", value = "value", x:y_adjusted_pubs) %>%
   group_by(variable) %>%
   mutate(value = value/max(value)) %>%
   ungroup() %>%
   spread(key = "variable", value = "value")

 return(data_scaled)
```

```
}

# run function
uni_data_scaled <- scale_all_vars(data_for_scaling)

glimpse(uni_data_scaled)
## Rows: 4
## Columns: 5
## $ institution        <chr> "University A", "University B",
"University C", "Un...
## $ year               <dbl> 2011, 2011, 2011, 2011
## $ x                  <dbl> 1.0000000, 0.4191176, 0.8529412, 0.3069853
## $ y_adjusted_load    <dbl>  0.9159292,  0.4407080,  1.0000000,
0.5584440
## $ y_adjusted_pubs    <dbl>  1.0000000,  0.7778182,  0.6969091,
0.3852727
```

8.4.4   Find Absolute Productivity for the Base Year

Now calculate 2011 absolute productivity. And just as in the introductory section, we'll replicate the data for the five time points observed in our dataset. Recall from above that we will multiply base absolute productivity scores with the cumulative productivity change scores for each subsequent year. Next create the function for doing this, and run it once for a 50-50 education-research value weighting scheme.

```
# set function parameters
abs_prod_calc <- function(df,
                              edu_weight = 0.5,
                              res_weight = 0.5,
                              edu_var = "y_adjusted_load",
                              res_var = "y_adjusted_pubs") {

  # calculate productivity from scaled variables
    abs_prod_base <- df %>%
                   mutate(y    =    edu_weight*y_adjusted_load    +
res_weight*y_adjusted_pubs,
          P = y / x) %>%
  select(-y_adjusted_pubs, -y_adjusted_load)

  # specify subsequent years
    subsequent_years <- c(2011, 2013, 2015, 2017, 2019)

  # replicate data for subsequent years
    make_data_year <- function(yr) {
```

```
      data <- abs_prod_base %>%
   mutate(year = yr)

      return(data)
   }

   # append replicated data
   abs_prod <- map_dfr(subsequent_years, make_data_year)

   # long format and specify weighting scheme
   abs_prod_repeated <- abs_prod %>%
      pivot_longer(x:P, names_to = "nature", values_to = "abs_base_
value") %>%
      arrange(institution, nature, year) %>%
       mutate(nature = ifelse(nature == "P",

                                         glue("{nature}:
e{edu_weight*100}-r{res_weight*100}"),
                        nature))

   return(abs_prod_repeated)
}

# run function for default weights
abs_prod_base_data <- abs_prod_calc(uni_data_scaled)

# check results
glimpse(abs_prod_base_data)
## Rows: 60
## Columns: 4
## $ institution <chr> "University A", "University A", "University
A", "Uni...
## $ year  <dbl> 2011,  2013,  2015,  2017,  2019,  2011,  2013,
2015, 2017...
## $ nature <chr> "P: e50-r50", "P: e50-r50", "P: e50-r50", "P:
e50-r5...
## $ abs_base_value  <dbl>  0.9579646,  0.9579646,  0.9579646,
0.9579646, 0.957964...
```

A quick look at the data tells you that we have our base-level 2011 productivity scores, along with x and y indexes for each university, repeated for each year.

8.4.5   Threshold Analysis

The next step is to test productivity scores across our value weight ranges. Just as
we did in Chap. 5, we'll use the function we created above to iterate over several
combinations of education and research values weights.

```
# create value weight ranges
edu_vals <- seq(from = 0.3, to = 0.7, by = 0.05)
res_vals <- seq(from = 0.7, to = 0.3, by = -0.05)

# iterate
abs_prod_range_base <- map2_dfr(edu_vals, res_vals, abs_prod_calc,
df = uni_data_scaled)

# check results
glimpse(abs_prod_range_base)
## Rows: 540
## Columns: 4
## $ institution <chr> "University A", "University A", "University
A", "Uni...
## $ year <dbl> 2011, 2013, 2015, 2017, 2019, 2011, 2013,
2015, 2017...
## $ nature <chr> "P: e30-r70", "P: e30-r70", "P: e30-r70", "P:
e30-r7...
## $ abs_base_value <dbl> 0.9747788, 0.9747788, 0.9747788,
0.9747788, 0.974778...
```

We can see now that we have far more observations in this data set because it's
created the base level productivity scores for each combination of education and
research value weights created above.

Now take this data and combine with the full results of our productivity change
threshold analysis from Chap. 5. Then multiply to get an alternative view of the data
that allows for more direct institution-to-institution comparisons of changes in
absolute productivity across the period.

```
# combine and multiply
abs_prod_range <- prod_range %>%
    bind_cols(abs_prod_range_base[,4]) %>%
    mutate(abs_prod = cum_change_composite*abs_base_value)

# check results
glimpse(abs_prod_range)
## Rows: 540
## Columns: 7
```

```
##   $   institution   <chr>   "University   A",   "University   A",
"University A"...
## $ year <dbl> 2011, 2013, 2015, 2017, 2019, 2011, 2013, 2015...
## $ nature <chr> "P: e30-r70", "P: e30-r70", "P: e30-r70", "P: ...
##   $   composite_index   <dbl>   1.0000000,   1.1018604,   1.0739799,
1.0303713, 1....
##   $   cum_change_composite   <dbl>   1.000000,   1.101860,   1.183376,
1.219317, 1.3524...
##   $   abs_base_value   <dbl>   0.9747788,   0.9747788,   0.9747788,
0.9747788, 0....
## $ abs_prod <dbl> 0.9747788, 1.0740701, 1.1535297, 1.1885639, 1....
```

We now have, side-by-side, our original range of productivity change portrayals
from Chap. 5 and our new range of absolute productivity portrayals. The next step
is to understand the significance and implications of both views of the data.

8.4.6   Plot the Data

Recall the plot_prod_range function we created in Chap. 5. We're going to take that
function and modify it, so that we can distinguish between the cumulative change
portrayal and the new absolute productivity portrayal.

```
# modify function parameters
plot_prod_range <- function(df,
                            prod_index,
                            var) {

  # specify different scales for the different indicators
  if(prod_index == "cum_change"){
    y_label <- "Cumulative change index"
  } else if (prod_index == "absolute"){
    y_label <- "Absolute productivity (normalised)"
  }

  # create plot
  plot <- df %>%
    filter(!nature %in% c("x", "y")) %>%
    ggplot(aes(x = year, y = .data[[var]], color = nature)) +
    facet_wrap(~institution) +
    geom_line() +
    geom_hline(yintercept = 1, linetype = "dashed") +
    scale_color_grey(start = 0.3, end = 0.8) +
    labs(x = "Year",
         y = y_label,
```

```
        color = "Weighting \nscheme",
        title = "Range of plausible productivity scores with \
nvariable education and research value weights")

    return(plot)
}


# overwrite function
save(plot_prod_range, file = "data/plot_prod_range")
```

Now recall the original cumulative productivity change thresholds where the reference for each institution's productivity change is only its own undefined level of productivity in the base year, and compare to the new absolute productivity thresholds (Figs. 8.2 and 8.3).

```
plot_prod_range(abs_prod_range,
                prod_index = "cum_change",
                var = "cum_change_composite")

plot_prod_range(abs_prod_range,
                prod_index = "absolute",
                var = "abs_prod")
```

We can see stark difference in the performance portrayals between the more traditional TI representations and the new absolute productivity change portrayals. First notice that Universities A and C started the period at a much lower base than

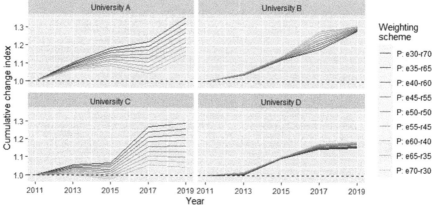

Fig. 8.2 Review of productivity change threshold analysis

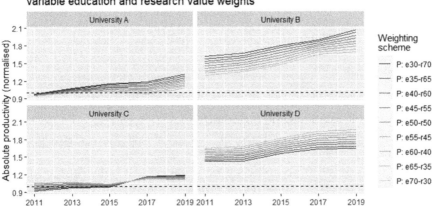

Fig. 8.3 Visualisation of absolute productivity thresholds

Universities B and D. Thus, although their research-emphasized performance portrayals are very complementary relative only to themselves, when you look at the absolute gains, the progress is not as positive.

A second new insight is the extent to which different treatments of education and research value weights have affected University B and D's initial starting points. Across the entire period, these universities heavily depend upon either a research-emphasized or an education-emphasized weighting scheme, respectively, for a more favorable portrayal. Further they seem to be holding on to a consistent strategy across the period. There is no apparent attempt to leverage new strengths in either education or research to bring their relative levels of performance in either category up to the level of the other.

8.4.7   Rankings Analysis

Let's now see what happens when we examine productivity change rankings with the absolute productivity scores across different value weights (Table 8.3 and 8.4).

```
get_ranks(abs_prod_range,
          var = "abs_prod")

get_ranks(abs_prod_range,
          yr = 2011,
          var = "abs_prod")
```

Table 8.3 Review of rankings based on productivity change

Institution	Year	P: e30-r70	P: e50-r50	P: e70-r30
University B	2019	1	1	2
University D	2019	2	2	1
University A	2019	3	3	4
University C	2019	4	4	3

Table 8.4 New absolute productivity change rankings

Institution	Year	P: e30-r70	P: e50-r50	P: e70-r30
University B	2011	1	2	2
University D	2011	2	1	1
University A	2011	3	4	4
University C	2011	4	3	3

As in Chap. 5, the value weighting system here also affects the university rankings. If all the universities were either robust or not robust to the different value weights in the same way, the rankings might not change very much. But the operations and strategies of different institutions are indeed very different, so subjectivity and values matter, not only in generating the more and less appropriate productivity portrayal for a single institution, but also they matter intensely for the field as a whole. When you start throwing more universities into the mix, and especially non-peer institutions, the picture becomes even more skewed.

8.5   Reflections on Absolute Productivity Threshold Analysis in Practice

The new absolute productivity change results from this chapter add some important clarity to the first productivity change results we produced in Chap. 5. They also served to flip our ideas of performance once again. When it comes to education institutions and NFPs, different treatments of data can drastically and dramatically change performance portrayals for single institutions and entire fields of organizations. This should serve as a clear lesson for any analyst or decision-maker.

Further, it's not that we are providing any evidence for unreliable data. All the mock data created in this book are closely based on real-world situations. On the contrary, we have developed some robust methods. The issues that emerge from all our opposing performance portrayals all stem from the different human values we have chosen to emphasize in each of our calculations.

If we attempt to measure the productivity of education institutions and NFPs, one of the most important tools we can bring to an analysis is an open mind. As we explore different ideas of performance, new valuable insights will inevitably emerge, and no single portrayal will necessarily be the "correct" one. No single analysis will ever stand on its own, and we should welcome surprising new results that challenge our initial assumptions.

Hypothesis Testing

9

9.1   Objective

This chapter demonstrates how to confirm and add credibility to the results of a productivity change analysis. By the end of the chapter, you will have experienced running a time series analysis on longitudinal productivity change data. You will experiment with two applications of running a regression with generalized least squares (GLS), a common technique used with time series data.

9.2   What Is Hypothesis Testing with Productivity Data?

Assume we have landed on an agreeable interpretation of productivity and an appropriate metric to capture it. There may remain questions about how reliable the observed trends are. As we have seen, not only can productivity data lend itself to different interpretations, but also it can be noisy. Inputs and outputs vary from one time point to the next, and depending on the purpose of the analysis, it may be important to confirm the reliability of measured results.

In this chapter, we'll pursue a simple question, has productivity changed or not? Our hypothesis is that productivity has changed meaningfully for the institutions over the period. The null hypothesis is that productivity has remained constant over the period, and any perceived change is more likely due to random chance or noise in the data.

Specifically, we want to know if productivity has changed because of some specific intervention. Below we'll briefly describe a scenario where, in 2011, a major policy change was enacted for all the universities under examination, and we want to know whether there exists evidence that the new policy has catalyzed changes in productivity for our four institutions.

© The Author(s), under exclusive license to Springer Nature
Switzerland AG 2021
K. Moore, *Measuring Productivity in Education and Not-for-Profits*,
Management for Professionals, https://doi.org/10.1007/978-3-030-72965-3_9

9.3   A Detailed Look at Hypothesis Testing
            for Productivity Change

9.3.1   Initial Caveat

In this chapter we'll conduct a time series analysis on our productivity data, but the
materials covered do not substitute for a true introduction to the topic. Time series
analysis represents its own extensive branch of data science. In this chapter we'll
use one illustrative technique to show how one might go about confirming the
results of the productivity model they've created. At the end of the chapter, we'll
discuss a few limitations of the approach.

9.3.2   Model Fitting with GLS

In the tutorial below, we're going to fit our observed productivity trend to a regression model using GLS. We use a GLS regression model because the value of each
subsequent data point from 2011 to 2019 to some extent depends upon the value of
the previous observation. We're looking at cumulative productivity change, so in
essence, a university can't achieve a score of 1.2 from 1.0 without at some point first
having reached a level of 1.1. This is a common phenomenon that must often be
accounted for in time series data, which an ordinary least squares (OLS) regression
does not correct for.

We'll be using the gls() function from the nlme package to run our regressions.
The gls() function has two important arguments that we'll need to specify regarding
how it operates on the data—the correlation argument and the method argument.
For the correlation argument of this function, we're going to specify corAR1(). We
use this argument when all the time intervals within our period are evenly spaced.
For the method argument, we're going to specify "ML," which stands for maximum
likelihood. We use this argument because the data we're examining do not represent
a random sample from a population.

In moving forward, we'll take those aspects of the regression for granted and
focus more on the procedural aspects of performing the overall analysis.

9.4   Tutorial: Hypothesis Testing Using Generalized
            Least Squares

9.4.1   Setup

```
# packages
library(knitr)
library(tidyverse)
library(nlme) # generalised linear models for time series
library(scales)
```

```
# data
load("data/prod_change_edu.rda")
```

9.4.2   Introduction

For some final context around our higher education scenario, assume that in 2011, the government capped its spending on subsidies for undergraduate tuition. The implication for institutions is that they now have to spread certain government grants more thinly across their growing student bodies. Some have suggested that the new funding policy has led to productivity increases for course delivery at many institutions. Others argue it hasn't made a difference.

To determine the policy effects, we're going to use the dataset we created in Chap. 4, prod_change_edu, that focused only on education productivity for our four universities. The 2011 policy had no direct implications for research.

9.4.3   Test Two Sets of Fitted Data

Our first step is to demonstrate how the hypothesis testing will work for a single institution. We're going to test whether University A has experienced a statistically significant change in education productivity over the period. The procedure we'll follow is to test whether there is any difference between a model where our y-values (productivity change) are dependent on our x-values (years) and a model where our y-values are fitted to a flat line at y = 1 (the null hypothesis). Well test this by running our models and then seeing if there is any statistically significant difference between them using an analysis of variance (ANOVA).

```
# filter only productivity scores for Uni A
uniA_prod_change <- prod_change_edu %>%
    filter(institution == "University A",
           nature == "P")

# Model 1 : Null hypothesis - fit the data to a flat line at y=1
model_1 <- gls(cum_change_composite ~ 1, cor = corAR1(), data =
uniA_prod_change, method = "ML")

# Model 2 : Linear fit for productivity cumulative change by year
model_2 <- gls(cum_change_composite ~ year, cor = corAR1(), data =
uniA_prod_change, method = "ML")

# check the difference between the two models
anova(model_2, model_1)$`p-value`[2]
## [1] 0.09165089
```

We can see that for University A, with a p-value of greater than 0.05, the change in productivity was not significantly different from a trend of no change across the period.

Now that we see how the process works, let's define a function that we can iterate over all universities to test whether their changes in productivity are statistically significant (Table 9.1). We'll also build into the function a feature that outputs the models from the GLS regression, so we can plot the fitted line against our data points.

```
# define function parameters
find_intervention_evidence <- function(df, inst,
                                        nat = "P",
                                        FUN = "test") {

  # filter for institution and productivity
  data <- df %>%
    filter(institution == inst,
           nature == nat)

  # specify percent change in productivity over the period
  diff <- data %>%
    filter(year == 2019) %>%
    mutate(percent_change = percent(cum_change_composite - 1)) %>%
    select(percent_change) %>%
    pull()

  # Model 1 : Null hypothesis -- flat line at y=1
  model_1 <- gls(cum_change_composite ~ 1, cor = corAR1(), data =
data, method = "ML")

  # Model 2 : Linear fit for productivity cumulative change by year
  model_2 <- gls(cum_change_composite ~ year, cor = corAR1(), data
= data, method = "ML")

  # communicate evidence for change
  if(FUN == "test"){
    test_result <- tibble(institution = inst,
                          productivity_percent_change = diff,
```

Table 9.1 Evidence of impact indicators

Institution	productivity_percent_change	p_value	Evidence
University A	1%	0.092	No evidence of impact
University B	33%	0.000	Evidence of impact
University C	−11%	0.001	Evidence of impact
University D	21%	0.000	Evidence of impact

```
                              p_value = round(anova(model_2, model_1)$`p-
value`[2], digits = 3),
                              evidence = ifelse(p_value > 0.05, "No
evidence of impact", "Evidence of impact"))
    return(test_result)

  # get predicted values for plotting
  } else if(FUN == "fit"){

    data_fitted <- data %>%
      mutate(predicted = predict(model_2))

    return(data_fitted)

  }
}
```

We can see that University A was the only institution to not experience a statistically significant change in productivity since the new policy was introduced.

Let's see what the fitted lines look like against our productivity data (Fig. 9.1).

```
# generate predicted values for plotting
fitted_data <- unique(prod_change_edu$institution) %>%
    map_dfr(find_intervention_evidence, df = prod_change_edu, FUN
= "fit")

# plot
ggplot(fitted_data, aes(x = year)) +
```

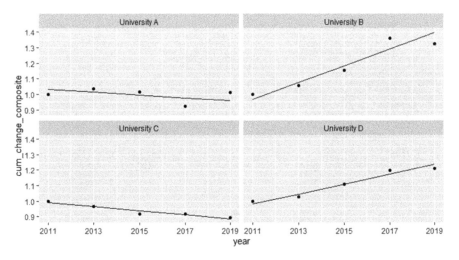

Fig. 9.1 Productivity change trends lines of best fit

```
geom_point(aes(y = cum_change_composite)) +
geom_line(aes(y = predicted)) +
facet_wrap(vars(institution)) +
scale_x_continuous(breaks = seq(2011, 2019, 2))
```

A quick glance at the data shows you that University A's change over the period is only very slight and generally a bit noisy, while the other's trends are most consistent.

9.4.4   Add Longitudinal Data

At this point, we still need to be critical. So three out of the four universities have experienced significant productivity changes since 2011, but what did their productivity trends look like before the policy change? Did the year 2011 change any trajectories, or were these universities headed in their respective directions anyway? University A hasn't changed much since the new policy, but might the new funding requirements have served to arrest a decline in productivity?

To answer these questions, we'll need to add some historical data.

```
# create data set
prior_prod_data <- tribble(
    ~institution,   ~year,  ~nature,  ~cum_change_composite,
    "University A",  2001,    "P",                   1.035,
    "University A",  2003,    "P",                   1.015,
    "University A",  2005,    "P",                   1.035,
    "University A",  2007,    "P",                   1.015,
    "University A",  2009,    "P",                   1.026,
    "University B",  2001,    "P",                   1.023,
    "University B",  2003,    "P",                   1.015,
    "University B",  2005,    "P",                   0.952,
    "University B",  2007,    "P",                   0.994,
    "University B",  2009,    "P",                   1.035,
    "University C",  2001,    "P",                   1.101,
    "University C",  2003,    "P",                   1.080,
    "University C",  2005,    "P",                   0.990,
    "University C",  2007,    "P",                   1.053,
    "University C",  2009,    "P",                   1.011,
    "University D",  2001,    "P",                   0.983,
    "University D",  2003,    "P",                   1.003,
    "University D",  2005,    "P",                   1.029,
    "University D",  2007,    "P",                   1.056,
    "University D",  2009,    "P",                   0.977
)
```

Now join the two data sets together, and add a dummy variable that signifies the periods before and after the policy was enacted.

```
# append new data to see extended time series
prod_change_extended <- prior_prod_data %>%
  bind_rows(prod_change_edu) %>%
  select(-composite_index) %>%
  filter(nature == "P") %>%
  mutate(policy_period = as.numeric(year >= 2011))
```

Now let's visualize the trends (Fig. 9.2).

```
# plot
ggplot(prod_change_extended, aes(x = year, y = cum_change_com-
posite)) +
  facet_wrap(~institution) +
  geom_line() +
  geom_point(data = filter(prod_change_extended, year == 2011),
size = 3) +
  geom_hline(yintercept = 1, linetype = "dashed") +
  xlab("Year") +
  ylab("Cumulative change") +
  theme(legend.position = "none")
```

A quick glance at Fig. 9.2 seems to imply that Universities A and C were not too affected, but B and D's trajectories appear as though they might have been influenced by the policy.

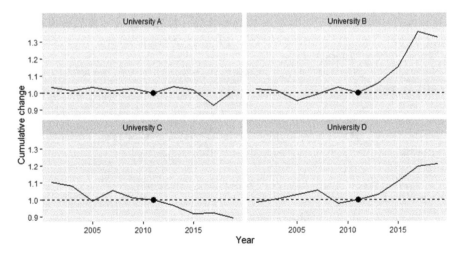

Fig. 9.2 Long-term productivity change trends

9.4.5   Did Our Intervention Make a Difference?

In order to further bolster our results, we need to see if there's a statistically significant interaction effect of the policy change. This time, our null hypothesis is that there is no interaction effect from the policy change, i.e., that the best fit of the data is just a continuous straight line fitted for our x and y values from 2001 to 2019. We want to see if there is any significant difference between that model of the data and one that actually fits two different lines to the data, i.e. one line for the data before 2011 and another line for the data after 2011. This kind of analysis is often called an interrupted time series analysis.

Let's replicate the function we created above, but with new GLS models serving the purpose described above.

```r
# define function parameters
find_evidence_long_term <- function(df, inst,
                                     nat = "P",
                                     FUN = "test") {

  # filter for institution and productivity
  data <- df %>%
    filter(institution == inst,
           nature == nat)

  # specify percent change in productivity over the period
  diff <- data %>%
    filter(year == 2019) %>%
    mutate(percent_change = percent(cum_change_composite - 1)) %>%
    select(percent_change) %>%
    pull()

  # Model 1 : null hypothesis: no interaction effect from policy
change -- continuous line
  model_1 <- gls(cum_change_composite ~ year, cor = corAR1(), data
= data, method = "ML")

  # Model 2 : policy period interaction effect
  model_2 <- gls(cum_change_composite ~ year * policy_period, cor
= corAR1(), data = data, method = "ML")

  # communicate evidence for interaction effect
  if(FUN == "test"){
    test_result <- tibble(institution = inst,
                          productivity_percent_change = diff,
                          p_value = round(anova(model_2, model_1)$`p-
value`[2], digits = 3),
```

```
                                        evidence = ifelse(p_value > 0.05, "No
evidence of impact", "Evidence of impact"))
    return(test_result)

  # # get predicted interaction effect values for plotting
  } else if(FUN == "fit"){

    data_fitted <- data %>%
      mutate(predicted = predict(model_2))

    return(data_fitted)

  }
}
```

Now run the function on our longitudinal data set (Table 9.2).

```
# run to find evidence for policy impact
evidence_long_term <- unique(prod_change_extended$institution) %>%
  map_dfr(find_evidence_long_term, df = prod_change_extended)

# see results
evidence_long_term
```

We can see that our initial assessment was correct. Only Universities B and D seem to have been impacted by the 2011 policy.

Let's plot the fitted data to see the trends pre and post the policy change (Fig. 9.3).

```
# get predicted values for plotting
fitted_data_long_term                 <-                unique(prod_change_
extended$institution) %>%
  map_dfr(find_evidence_long_term, df = prod_change_extended, FUN
= "fit")

# plot
ggplot(fitted_data_long_term, aes(x = year)) +
  geom_point(aes(y = cum_change_composite)) +
  geom_line(aes(y = predicted, group = policy_period)) +
```

Table 9.2 Evidence of impact for interaction effects

Institution	productivity_percent_change	p_value	Evidence
University A	1%	0.135	No evidence of impact
University B	33%	0.001	Evidence of impact
University C	−11%	0.632	No evidence of impact
University D	21%	0.002	Evidence of impact

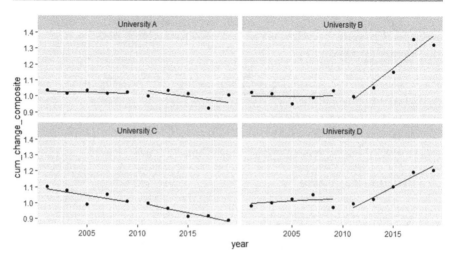

Fig. 9.3 Visualisation of pr- and post policy change trends

```
facet_wrap(vars(institution))
```

With the p-values and plots we've generated, we now have evidence that supports some of the initial assumptions about this policy effects, described above in the introduction section. A clear next step would be to conduct some case studies to find more detailed evidence about the individual institutions and investigate potentially what made some of them more sensitive to the policy change than others.

9.5   Reflections on Hypothesis Testing in Practice

It is first important to recognize the significance of having conducted both tests in this tutorial. If the prime assumption was that the policy should have led to productivity gains, the first test could directly answer that question. Any institution whose productivity did not show a statistically significant increase was likely not affected by the policy in its intended way.

It wasn't enough to stop there, however. The second test was necessary to build evidence that—regardless of the direction of the trend since 2011—the trend since the policy was enacted was different from the trend before the policy was enacted. Each test addresses a very different question, but both are necessary to build evidence for the impact of the policy.

Finally, it should be noted that the method we have chosen here may not be the best method for scaling up to an analysis of, say, a couple dozen individual institutions or more. If we hold ourselves to a p-value threshold of 0.05 for statistical significance, and we find that the results for several institutions are hovering around that mark, then we have to appreciate the true meaning of that p-value. Recall that it signifies the probability that our results are due to random chance, and 0.05 is just

1/20. So, if we are performing two tests each for 20 or more institutions, then it's not unreasonable that at least one or more of our statistically significant results is just a fluke.

Exploring alternative time series methods could be one way to address the problem. Another solution might be simply to aggregate multiple institution's data together. There will always be caveats around any route you take, but hopefully this has provided a firm launching pad for you to conduct some compelling productivity analyses on our own data.

Index

© The Author(s), under exclusive license to Springer Nature
Switzerland AG 2021
K. Moore, *Measuring Productivity in Education and Not-for-Profits*,
Management for Professionals, https://doi.org/10.1007/978-3-030-72965-3

Lightning Source UK Ltd.
Milton Keynes UK
UKHW020816080722
405567UK00002B/4

9 783030 729677